ESTIMATING MORTALITY RISK REDUCTION AND ECONOMIC BENEFITS FROM CONTROLLING OZONE AIR POLLUTION

Committee on Estimating Mortality Risk Reduction Benefits from Decreasing Tropospheric Ozone Exposure

Board on Environmental Studies and Toxicology

Division on Earth and Life Studies

NATIONAL RESEARCH COUNCIL
OF THE NATIONAL ACADEMIES

THE NATIONAL ACADEMIES PRESS
Washington, D.C.
www.nap.edu

THE NATIONAL ACADEMIES PRESS 500 Fifth Street, NW Washington, DC 20001

NOTICE: The project that is the subject of this report was approved by the Governing Board of the National Research Council, whose members are drawn from the councils of the National Academy of Sciences, the National Academy of Engineering, and the Institute of Medicine. The members of the committee responsible for the report were chosen for their special competences and with regard for appropriate balance.

This project was supported by Contract 68-C-03-081 between the National Academy of Sciences and the U.S. Environmental Protection Agency. Any opinions, findings, conclusions, or recommendations expressed in this publication are those of the authors and do not necessarily reflect the view of the organizations or agencies that provided support for this project.

International Standard Book Number-13: 978-0-309-11994-8
International Standard Book Number-10: 0-309-11994-4

Additional copies of this report are available from

The National Academies Press
500 Fifth Street, NW
Box 285
Washington, DC 20055

800-624-6242
202-334-3313 (in the Washington metropolitan area)
http://www.nap.edu

Printed in the United States of America.

THE NATIONAL ACADEMIES

Advisers to the Nation on Science, Engineering, and Medicine

The **National Academy of Sciences** is a private, nonprofit, self-perpetuating society of distinguished scholars engaged in scientific and engineering research, dedicated to the furtherance of science and technology and to their use for the general welfare. Upon the authority of the charter granted to it by the Congress in 1863, the Academy has a mandate that requires it to advise the federal government on scientific and technical matters. Dr. Ralph J. Cicerone is president of the National Academy of Sciences.

The **National Academy of Engineering** was established in 1964, under the charter of the National Academy of Sciences, as a parallel organization of outstanding engineers. It is autonomous in its administration and in the selection of its members, sharing with the National Academy of Sciences the responsibility for advising the federal government. The National Academy of Engineering also sponsors engineering programs aimed at meeting national needs, encourages education and research, and recognizes the superior achievements of engineers. Dr. Charles M. Vest is president of the National Academy of Engineering.

The **Institute of Medicine** was established in 1970 by the National Academy of Sciences to secure the services of eminent members of appropriate professions in the examination of policy matters pertaining to the health of the public. The Institute acts under the responsibility given to the National Academy of Sciences by its congressional charter to be an adviser to the federal government and, upon its own initiative, to identify issues of medical care, research, and education. Dr. Harvey V. Fineberg is president of the Institute of Medicine.

The **National Research Council** was organized by the National Academy of Sciences in 1916 to associate the broad community of science and technology with the Academy's purposes of furthering knowledge and advising the federal government. Functioning in accordance with general policies determined by the Academy, the Council has become the principal operating agency of both the National Academy of Sciences and the National Academy of Engineering in providing services to the government, the public, and the scientific and engineering communities. The Council is administered jointly by both Academies and the Institute of Medicine. Dr. Ralph J. Cicerone and Dr. Charles M. Vest are chair and vice chair, respectively, of the National Research Council.

www.national-academies.org

COMMITTEE ON ESTIMATING MORTALITY RISK REDUCTION BENEFITS FROM DECREASING TROPOSPHERIC OZONE EXPOSURE

Members

JOHN C. BAILAR III (*Chair*), University of Chicago, IL
RICHARD T. BURNETT, Health Canada, Ottawa, ON, Canada
LAURAINE G. CHESTNUT, Stratus Consulting Inc., Boulder, CO
W. MICHAEL FOSTER, Duke University Medical Center, Durham, NC
A. MYRICK FREEMAN, III, Bowdoin College, Brunswick, ME
MONTSERRAT FUENTES, North Carolina State University, Raleigh, NC
DANIEL S. GREENBAUM, Health Effects Institute, Boston, MA
ALAN KRUPNICK, Resources for the Future, Washington, DC
NINO KÜNZLI, Center for Research in Environmental Epidemiology at Municipal Institute of Medical Research, Barcelona, Spain
KENT E. PINKERTON, University of California, Davis, CA
ARMISTEAD G. RUSSELL, Georgia Institute of Technology, Atlanta, GA
HELEN SUH, Harvard School of Public Health, Boston, MA
EVELYN O. TALBOTT, University of Pittsburgh Graduate School of Public Health, PA

Staff

RAYMOND A. WASSEL, Project Director
NORMAN GROSSBLATT, Senior Editor
MIRSADA KARALIC-LONCAREVIC, Manager, Technical Information Center
JOHN BROWN, Program Associate
RADIAH ROSE, Senior Editorial Assistant

Sponsor

U.S. ENVIRONMENTAL PROTECTION AGENCY

---

[1]This study was planned, overseen, and supported by the Board on Environmental Studies and Toxicology.

OTHER REPORTS OF THE BOARD ON ENVIRONMENTAL STUDIES AND TOXICOLOGY

Respiratory Disease Research at NIOSH (2008)
Evaluating Research Efficiency in the U.S. Environmental Protection Agency (2008)
Hydrology, Ecology, and Fishes of the Klamath River Basin (2008)
Applications of Toxicogenomic Technologies to Predictive Toxicology and Risk Assessment (2007)
Models in Environmental Regulatory Decision Making (2007)
Toxicity Testing in the Twenty-first Century: A Vision and a Strategy (2007)
Sediment Dredging at Superfund Megasites: Assessing the Effectiveness (2007)
Environmental Impacts of Wind-Energy Projects (2007)
Scientific Review of the Proposed Risk Assessment Bulletin from the Office of Management and Budget (2007)
Assessing the Human Health Risks of Trichloroethylene: Key Scientific Issues (2006)
New Source Review for Stationary Sources of Air Pollution (2006)
Human Biomonitoring for Environmental Chemicals (2006)
Health Risks from Dioxin and Related Compounds: Evaluation of the EPA Reassessment (2006)
Fluoride in Drinking Water: A Scientific Review of EPA's Standards (2006)
State and Federal Standards for Mobile-Source Emissions (2006)
Superfund and Mining Megasites—Lessons from the Coeur d'Alene River Basin (2005)
Health Implications of Perchlorate Ingestion (2005)
Air Quality Management in the United States (2004)
Endangered and Threatened Species of the Platte River (2004)
Atlantic Salmon in Maine (2004)
Endangered and Threatened Fishes in the Klamath River Basin (2004)
Cumulative Environmental Effects of Alaska North Slope Oil and Gas Development (2003)
Estimating the Public Health Benefits of Proposed Air Pollution Regulations (2002)
Biosolids Applied to Land: Advancing Standards and Practices (2002)
The Airliner Cabin Environment and Health of Passengers and Crew (2002)
Arsenic in Drinking Water: 2001 Update (2001)
Evaluating Vehicle Emissions Inspection and Maintenance Programs (2001)
Compensating for Wetland Losses Under the Clean Water Act (2001)
A Risk-Management Strategy for PCB-Contaminated Sediments (2001)
Acute Exposure Guideline Levels for Selected Airborne Chemicals (five volumes, 2000-2007)
Toxicological Effects of Methylmercury (2000)
Strengthening Science at the U.S. Environmental Protection Agency (2000)
Scientific Frontiers in Developmental Toxicology and Risk Assessment (2000)
Ecological Indicators for the Nation (2000)

Waste Incineration and Public Health (2000)
Hormonally Active Agents in the Environment (1999)
Research Priorities for Airborne Particulate Matter (four volumes, 1998-2004)
The National Research Council's Committee on Toxicology: The First 50 Years (1997)
Carcinogens and Anticarcinogens in the Human Diet (1996)
Upstream: Salmon and Society in the Pacific Northwest (1996)
Science and the Endangered Species Act (1995)
Wetlands: Characteristics and Boundaries (1995)
Biologic Markers (five volumes, 1989-1995)
Review of EPA's Environmental Monitoring and Assessment Program (three volumes, 1994-1995)
Science and Judgment in Risk Assessment (1994)
Pesticides in the Diets of Infants and Children (1993)
Dolphins and the Tuna Industry (1992)
Science and the National Parks (1992)
Human Exposure Assessment for Airborne Pollutants (1991)
Rethinking the Ozone Problem in Urban and Regional Air Pollution (1991)
Decline of the Sea Turtles (1990)

Copies of these reports may be ordered from the National Academies Press
(800) 624-6242 or (202) 334-3313
www.nap.edu

Preface

The U.S. Environmental Protection Agency (EPA) asked for an independent study by the National Research Council to evaluate the scientific and technical bases of approaches used by EPA for estimating ozone-mortality reduction and associated benefits of health-based standards over time. In response, the National Research Council established the Committee on Mortality Risk Reduction Benefits from Decreasing Tropospheric Ozone Exposure. Biographic information on the committee members is presented in Appendix A.

In the course of preparing this report, the committee met four times. At three of the meetings—which were held in Irvine, CA; Washington, DC; and Woods Hole, MA—officials of EPA and academic researchers were invited to meet with the committee and present their views and results of their work. Interested members of the public at large were also given an opportunity to speak on those occasions. The fourth meeting was held in closed session so that the committee could complete drafting its report. Subsequently, the committee held three teleconferences to complete its deliberations.

As this report was being written, EPA was in the process of reviewing the existing National Ambient Air Quality Standards (NAAQS) for ozone. The primary (health-based) ozone NAAQS was set at 0.08 parts ppm for the annual fourth-highest daily maximum 8-h average concentration, averaged over 3 years. The committee's statement of task (see Chapter 1) and its deliberations were not dependent on EPA's decisions provided in its final rule on March 12, 2008 which lowered the level of the 8-h standard to 0.075 ppm.

The committee received oral and written presentations from John Balmes, University of California, San Francisco; Michelle Bell, Yale University; Kiros Berhane, University of Southern California; J.R. DeShazo, University of California, Los Angeles; James Hammitt, Harvard University; Bryan Hubbell, EPA; Al McGartland, EPA; Joel Schwartz, Harvard University; Anne Smith, CRA International; Deborah Shprentz, consultant to the American Lung Association; and Ira Tager, University of California, Berkeley.

Nathalie Simon, of EPA, provided the committee with information from EPA and the published scientific literature.

This report has been reviewed in draft form by persons chosen for their diverse perspectives and technical expertise in accordance with procedures approved by the National Research Council's Report Review Committee. The purpose of the independent review is to provide candid and critical comments that will assist the institution in making its published report as sound as possible and to ensure that the report meets institutional standards of objectivity, evidence, and responsiveness to the study charge. The review comments and draft manuscript remain confidential to protect the integrity of the deliberative process. We thank the following for their review of this report: Francesca Dominici, Johns Hopkins University; Mark Frampton, University of Rochester; John Graham, Frederick S. Pardee RAND Graduate School; Jane Hall, California State University; James Hammitt, Harvard University; Fintan Hurley, Institute of Occupational Medicine (in the United Kingdom); Jonathan Levy, Harvard University; Thomas Lumley, University of Washington; Frederick Lurmann, Sonoma Technology, Inc.; Jennifer Peel, Colorado State University; Richard Smith, University of North Carolina; Ira Tager, University of California, Berkeley; and Sverre Vedal, University of Washington.

Although the reviewers listed above have provided many constructive comments and suggestions, they were not asked to endorse the conclusions or recommendations, nor did they see the final draft of the report before its release. The review of the report was overseen by Gilbert Omenn, University of Michigan Medical School, and Edwin Clark, II, Earth Policy Institute. Appointed by the National Research Council, they were responsible for making certain that an independent examination of the report was carried out in accordance with institutional procedures and that all review comments were carefully considered. Responsibility for the final content of the report rests entirely with the author committee and the institution. We thank the report review monitor and coordinator.

We also thank Bailus Walker, Jr. for his constructive service on the committee; he resigned from the committee for personal reasons.

The committee's work for this report was assisted by staff of the National Research Council's Board on Environmental Studies and Toxicology (BEST). We thank Raymond Wassel, project director, and James Reisa, director of BEST. Technical information was provided by Mirsada Karalic-Loncarevic. Invaluable logistical support was provided by John Brown. Other staff members who contributed to this effort are Radiah Rose (senior editorial assistant) and Heidi Murray-Smith (research associate). The report was edited by Norman Grossblatt.

John C. Bailar III, *Chair*
Committee on Mortality Risk Reduction
Benefits from Decreasing Tropospheric
Ozone Exposure

Contents

BOXES, FIGURES, AND TABLES

BOXES

FIGURES

TABLES

ESTIMATING MORTALITY RISK REDUCTION AND ECONOMIC BENEFITS FROM CONTROLLING OZONE AIR POLLUTION

Summary

Increased concentrations of ozone in the lower atmosphere are formed from pollutants emitted by such human activities as the combustion of fossil fuels. Natural sources of emissions, such as vegetation, also contribute to ozone formation. Because human exposure to ozone in the lower atmosphere at the increased concentrations that result from precursor emissions can cause respiratory problems and other health effects, ozone is one of the six criteria pollutants regulated by the U.S. Environmental Protection Agency (EPA) under the Clean Air Act.[1]

Studies published since 1990 have yielded mixed evidence of a relationship between short-term exposure to ozone and premature death. According to analyses of evidence published in recent years, the risk of death in the population increases slightly but consistently as exposure to ozone increases. However, at the same time, interpretation of this evidence by EPA and scientists outside the agency has been complicated by ozone's occurrence in mixtures with other pollutants that have similar effects and by uncertainties including those which result from using outdoor-ozone measurements to estimate exposures of people who spend most of their time indoors.

The Clean Air Act requires EPA to review periodically the National Ambient Air Quality Standards (NAAQS) for the criteria pollutants.[2] Each time NAAQS are reviewed, the EPA administrator must weigh the most recent evidence and current uncertainties and make a public-health policy judgment about whether the existing standards are adequate to protect the public health with an adequate margin of safety or should be lowered or raised.

---

[1]Most ozone in the lower atmosphere is formed by a complex series of photochemical reactions in the presence of sunlight involving nitrogen oxides and volatile organic compounds. *Ozone* is used here to refer to the broad array of photochemical oxidants in ambient air, of which ozone is the primary component.

[2]As this report was being prepared, EPA was reviewing the NAAQS for ozone. EPA's final decision on the ozone NAAQS based on its review was announced in March 2008.

After the NAAQS are determined, EPA must address any mitigation measures needed to reduce emissions. When deciding on mitigation actions expected to cost more than $100 million per year, EPA, like other federal agencies, is required to carry out a cost-benefit analysis of alternative regulatory strategies, such as those to attain the ozone NAAQS. However, EPA is not allowed to consider monetary costs when setting NAAQS.

To assess the benefits portion of its cost-benefit analysis of the ozone NAAQS, EPA uses results of epidemiologic studies to estimate the number of premature deaths avoided by the expected reduction in ozone concentration for the population at risk (that is, the number of deaths postponed to some future year and generally with a different cause of death). It then assigns a monetary value to the avoided deaths by using a concept known as the value of a statistical life. That value is derived from studies of adults (mostly of working age) who indicate or reveal, through choices in the labor market or in other ways, the amount that they would be willing to pay to change their risk of death in a given period by a small amount. EPA applies the same value to all premature deaths avoided regardless of the age or health status of the population experiencing the potential change in risk of death. However, the willingness to pay for a reduction in the risk of death hypothetically depends on the characteristics (for example, life expectancy or health status) of the individuals affected or on the nature of the risk (for example, accident vs illness).

In light of recent evidence on the relationship of ozone to mortality and questions about its implications for benefit analysis, EPA asked the National Research Council to establish a committee of experts to evaluate independently the contributions of recent epidemiologic studies to understanding the size of the ozone-mortality effect in the context of benefit analysis. The committee was also asked to assess methods for estimating how much a reduction in short-term exposure to ozone would reduce premature deaths, to assess methods for estimating associated increases in life expectancy, and to assess methods for estimating the monetary value of the reduced risk of premature death and increased life expectancy in the context of health-benefits analysis. The charge to this National Research Council committee focused on benefit analysis; it did not include considering how evidence is used to set the ozone NAAQS.[3]

OVERALL CONCLUSIONS AND RECOMMENDATIONS

The committee concludes from its review of the health-based evidence that short-term exposure to ambient ozone is likely to contribute to premature deaths. Despite some continuing questions about the interpretation of the evidence, the committee concluded that the evidence is strong enough to be used in the estimation of the expected mortality-reduction benefits of a decrease in exposure to ozone. Human chamber and toxicologic studies have yielded strong

[3]A full statement of the committee's charge is presented in Box 1-1 of Chapter 1.

evidence indicating that short-term exposure to ozone can exacerbate lung conditions, causing illness and hospitalization, and can potentially lead to death. The available evidence on ozone exposure and exacerbation of heart conditions, which is less abundant, points to another concern. Epidemiologic studies have also found that exposure to ozone is associated with adverse lung and heart effects.

> **Recommendation: The committee recommends that ozone-related mortality be included in future estimates of the health benefits of reducing ozone exposure. The committee further recommends that the greatest emphasis be placed on estimates from new systematic multicity analyses that use national databases of air pollution and mortality, such as in the National Morbidity, Mortality, and Air Pollution Study, without excluding consideration of meta-analyses of previously published studies. Emphasis should also be placed on risk estimation based on analyzing data on multiple days so that delayed acute effects estimates can be included. The health-benefits estimates should be accompanied by a broad array of analyses of uncertainty but should give little or no weight to the assumption that there is no causal association between estimated reductions in premature mortality and reduced ozone exposure.**

Because older persons appear to be at higher risk of health-related effects from ozone pollution, it is appropriate to consider whether the willingness to pay for mortality risk reductions should and could reflect the number of years of life by which life would be extended by reductions in ozone. **The committee concludes that the evidence is insufficient to support a specific adjustment of the aggregate willingness to pay for reduction in annual mortality risk on the basis of differences in remaining life expectancy.**

> **Recommendation: Although there are many concerns about the accuracy of a willingness-to-pay (WTP) value and the corresponding value of a statistical life (VSL) that does not vary with population or risk characteristics, the committee recommends the use of a constant WTP and corresponding VSL as the most scientifically supportable approach to monetary valuation of ozone-related mortality risk given the information available in the epidemiologic and economics literature.**

INTERPRETATION OF RESULTS OF HEALTH STUDIES

The associations between ozone exposure and premature mortality in the recent health studies appear robust, but several factors create considerable uncertainty about them. Those factors can affect estimates of risk of ozone-related mortality in various ways. In some cases, the factors would cause an underestimation of risk; in other cases, an overestimation. On balance, the committee considers the evidence from the studies to be strong enough for use in deriving risk estimates, but the various factors and their potential effects on the estimates should be fully acknowledged.

Short-Term Exposure to Ambient Ozone

Time-series epidemiologic studies of short-term effects of ozone typically characterize human exposure by using ambient concentrations measured at fixed outdoor monitoring sites. Exposure is characterized by applying an averaging period to the daily ambient monitoring data. Changes in the average values are then linked with changes in mortality. When averaged over 24 h, ambient concentrations are weakly associated with corresponding personal ozone exposure, although the association is stronger in the summer than in winter. For shorter averaging periods, such as the afternoon (when both personal outdoor activity and ozone concentration can be at their highest), results from one study suggest that hourly or daily peak ambient ozone concentration may be an appropriate proxy for corresponding hourly or peak personal exposure. Whether observations from that study are relevant for people at risk for ozone-related death warrants further examination.

The choice of averaging period to characterize short-term ozone exposure in linking ambient ozone concentration with mortality risk can have a large effect on estimates of benefits of emission-control programs. For example, under some conditions, efforts to lower emissions of oxides of nitrogen could reduce the daily peak ozone concentration but raise daily average concentrations (see Chapter 3). Thus, a cost-benefit analysis of an emission-control program that examines mortality and daily average concentrations of ozone could appear to have a negative effect, whereas an analysis that examines mortality and daily peak concentrations could appear to show a benefit. It is not known which averaging period reflects personal exposure more accurately or is more closely related to mortality risk.

Recommendation: Future studies of the effects of short-term ozone exposure should determine whether and how much daily peak exposures, such as 1-h or 8-h exposures, and longer-term average exposures, such as over 24 h, are associated with ozone-related mortality. Benefits assessors at EPA and elsewhere should use the results to identify the appropriate exposure averaging periods so that they can estimate how efforts to attain the ozone NAAQS will affect ozone exposure and health. Regulators should take into account the possibility that the effects of ozone-control strategies averaged over 24 h may be quite different from those averaged over shorter periods.

Potential Confounding by Other Pollutants

Studies have not been sufficient to control fully for potential confounding by or ozone interaction with constituents of airborne particulate matter that has a diameter equal to or less than 2.5 μm ($PM_{2.5}$). Such constituents include sulfates, acids, elemental and organic carbon, and metals. The potential for confounding of ozone health effects by $PM_{2.5}$ constituents differs by region and season. For example, in the eastern United States, confounding is most likely in the summer months, when ozone and $PM_{2.5}$ are strongly correlated in many locations. In the winter, the potential for confounding is likely to be less. It will be difficult to address such confounding with currently available data, however, because data on $PM_{2.5}$ components at many sites have only recently begun to be routinely collected and because winter ozone concentrations (see next section) are often not measured.

Recommendation: Epidemiologic research on associations between air pollution and health outcomes should investigate regional and seasonal associations between ambient (outdoor) concentrations and human exposures to ozone and $PM_{2.5}$. It should also investigate how data on pollutants can be used to control for confounding and correlations between the various pollutant measures. When possible, researchers should address those issues by focusing on groups of people who are sensitive to ozone and by using data on the chemical and physical components and size distribution of ambient particles.

Recommendation: EPA and the scientific community should account for seasonal and geographic variability in the relationship between ozone and its potential confounders and should increasingly include the growing database on $PM_{2.5}$ constituents in analyses of confounding of ozone associations. The most relevant particle-component data should be collected frequently enough to improve understanding of the potential for confounding.

Ozone-Mortality Relationships During Winter Months

There is a lack of observed association between ozone and mortality during periods when ozone is low, such as winter. Reasons for the lack of association are not well understood in part because of the decrease in monitoring during those periods. Better understanding of ozone-mortality relationships in the winter is important for full exploration of effects at low concentrations. Although ambient ozone is one of the best-characterized pollutants in the United States, ozone monitors are usually operated only during the so-called ozone season—the warmer period of the year, which varies from city to city.

Recommendation: EPA and states should extend operation of ozone monitoring into winter and report the results. The winter program should be sufficient to allow researchers to examine seasonal differences in risk, how these seasonal differences vary spatially between communities with warmer and cooler winters, and ozone-mortality relationships at lower ozone concentrations. Ozone is a regional pollutant, so winter measurements need not be collected at all the summer locations; but if measurements are collected in winter, they should be collected with the same frequency as summer measurements.

Frailty and Ozone Mortality

Benefits assessors seek information on whether the mortality risk associated with acute ozone exposure is attributable to short-term displacement (in this case, advancement) of deaths that would have occurred in a few days or more without acute ozone exposure. If it is so attributable, they can focus their efforts on estimating the value that frail people would place on reducing their ozone-mortality risk.

On the basis of available evidence, the committee concludes that deaths related to exposure to ozone are not restricted to people who are at high risk of death within a few days. For example, a recent study of data collected from several U.S. cities reports that short-term ozone exposure is likely to contribute to shortening life and not only among people who are near death. However, because the evidence comes from only one study, it warrants confirmation by other studies.

Recommendation: EPA and the scientific community should conduct additional studies to investigate whether and how much ozone-related mortality is restricted to people who are already at high risk of death within a few days and how much ozone-related mortality occurs in people who are not already at risk of death in a few days. The studies should include use of various methods, for example, focus on investigating subjects who have diseases, such as diabetes or heart disease, that are known to be associated with air-pollution-related mortality risk.

Susceptibility

Preliminary results indicate that the effect of acute ozone exposure on mortality is likely to be larger than average in persons with pre-existing disease, especially lung and heart diseases. The list of factors that plausibly modify effects is rather long and still insufficiently investigated (see Chapter 4). Although susceptibility factors are important, the distribution of ozone-mortality-effect estimates among the categories of susceptibility is not adequately known. Consequently, the overall (population-weighted average) mortality effect in the total population is the only currently available basis of risk assessments; this approach is a source of an unknown amount of uncertainty in calculating the overall benefit of an ozone reduction in reducing mortality risk.

Recommendation: EPA and the scientific community should identify personal characteristics that are important in understanding ozone-mortality relationships. They should develop a distribution of ozone-mortality effect estimates among the categories of susceptibility; this will enable benefits assessment to include quantitative details of the heterogeneity of effects in the total population.

Presence or Absence of a Threshold for Ozone-Related Mortality

The association between short-term variations in ambient ozone concentrations and fluctuations in mortality rates is usually characterized as linear. Because the association is based on epidemiologic studies that can only approximate exposure on the basis of ambient monitoring data, the assumption of linearity should be viewed with caution. At low ozone concentrations, the question arises whether the association is linear or more accurately characterized as having a "threshold"—a concentration of ozone below which exposure poses no risk of death. Individuals have their own susceptibility, which is characterized by a unique exposure-response association; this association may include a unique threshold value that can vary with susceptibility of the individual at any given moment and with the averaging period used to assess exposure.

On the basis of its review of the evidence, the committee concludes that the association between short-term changes in ozone concentrations and mortality is generally linear throughout most of the concentration range, although uncertainties make it difficult to determine whether there is a threshold for the association at the lower end of the range. If there is a threshold, it is likely to be below the current NAAQS.

Recommendation: EPA and the scientific community should explore further how personal thresholds may vary and the extent to which one's threshold depends on one's frailty at any given moment. Because it is not clear whether there is an association between ozone and mortality in the cooler months, warmer months should be examined separately. The research should involve panel studies of individuals considered to be susceptible to premature death from ozone exposure, such as those with impaired lung or heart function. A sensitivity analysis is needed to assess how different thresholds in exposure-response relationships may affect ozone-mortality risk estimates that are based on results of epidemiologic studies.

Accounting for a Lag in Mortality Response to Ozone

Deaths related to short-term ozone exposure may not occur until several days after the exposure or may be associated with multiple short-term exposures. Many studies of short-term effects investigate a change in death rates for only one or a few days, but distributed-lag models can be used to look further ahead to capture delayed mortality, often referred to as a subacute response. **Distributed-lag analyses appear to capture the overall effects of ozone better than do single-day models, but there have been relatively few such analyses.**

> **Recommendation: EPA and the scientific community should develop appropriate databases and conduct distributed-lag analyses in future epidemiologic investigations to improve understanding of the statistical distribution of time between an increase in the ambient ozone concentration and the occurrence of deaths.**

Chronic Exposures

EPA benefits assessments have not included estimates of mortality risk due to long-term (chronic) exposure to ozone, because evidence does not directly demonstrate a causal relationship when the period between exposure and death is longer than a few days. However, the observed associations between ozone exposure and decreased small-airway lung function during childhood and adolescence suggest that ozone-related mortality is at least partially attributable to exposures that last more than a few days. The general association between poor lung function and shortened life expectancy is strong and well established, so evidence of an effect of ozone exposure on lung function increases the plausibility of mortality from chronic exposure. **The weak current evidence from cohort studies of an association of premature mortality with chronic exposure to ozone suggests that risks may be larger than those observed in acute effects studies alone.**

The standard approach to investigating effects of cumulative ozone exposure on life expectancy is the cohort study, in which large numbers of subjects are followed for several years. After taking into account all other factors that are likely to affect mortality, cohort studies can test the null hypothesis that mortality is the same among populations that have different ozone-exposure histories. However, none of the cohort studies available was designed to investigate chronic effects of ozone, and differences in ozone exposure among subjects in each study tended to be rather small. **If further confirmed, the weak current evidence from cohort studies of an association of premature mortality with longer-term exposure would support the notion that effects seen in time-series studies reflect only a portion of the total effect.**

Recommendation: EPA, the National Institutes of Health, and the scientific community should encourage additional studies of the association between long-term ozone exposure and mortality. They should also encourage development of models of long-term ozone exposures that can account for variations in exposure at the individual level and between and within cities. As new cohort evidence of effects of chronic exposures becomes available, EPA should consider including it in its benefits assessments.

Effect of Ozone Exposure on Life Expectancy

Effects of long-term cumulated exposure are, by design, not addressed in short-term time-series studies. Distributed-lag models integrate the distribution of the time between exposure and death, but they focus on a short window (several days to weeks) after exposure. **It is appropriate to use time-series results to estimate changes in life expectancy due to acute exposures by using cohort life-table methods if it is assumed that all members of the population at risk for death from ozone exposure have the same life expectancy as others in the same age and sex cohort (see Chapter 4). However, the committee finds that that assumption is questionable because people at greatest risk for death from short-term ozone exposure are likely to be those who have pre-existing diseases and thus life expectancy lower than average for their age.**

Recommendation: Additional studies are needed to assess the extent to which differences in susceptibility in a general population affect the variability associated with mortality risk estimates. To the extent that data are not available, models and assumptions can be used for sensitivity analysis to assess how risk estimates might vary with susceptibility.

Characterizing Mortality Risk by Using Studies of Acute and Chronic Exposure

Ozone-mortality risk is often expressed as the expected number of deaths attributable to ozone air pollution or lives saved by reducing ozone pollution by some amount. However, reductions in ozone exposure are expected to increase life expectancy and decrease age-adjusted annual death rates (for example, number of deaths per 100,000 of population). Thus, the number of older people would increase, the absolute number of deaths at higher ages would increase, and the annual number of deaths would return to normal in future years, although they would occur at higher ages and probably with different causes. Alternative approaches to the expression of ozone-mortality effects, such as death rates, will have similar results if one is concerned only with short-term effects of pollution changes. However, when one includes subacute and chronic effects estimates on mortality, the discrepancies between the results of the different approaches increase, and the conceptual flaws of the "attributable-cases" model become more pronounced.

Recommendation: EPA should evaluate alternative approaches for expressing ozone-mortality risk associations and consider the implications of using them in benefits assessments. EPA should consider placing greater emphasis on reporting changes in age-specific death rates in the relevant population and develop models for consistent calculation of changes in life expectancy and changes in numbers of deaths at all ages.

VALUATION

Willingness-to-Pay Estimates

Estimates of the value of a statistical life (VSL) are derived from estimates of an individual's willingness to pay (WTP) for changing his or her mortality risk by a small amount in a given period (usually annual) (see Chapter 2). The objective of an economic-benefits assessment is to develop an aggregate estimate of the welfare gain for everyone who benefits from a policy or program that is intended to reduce risk. Both economic theory and available empirical evidence are inconclusive about how people's WTP values vary with two important individual characteristics: age, as a proxy for remaining life expectancy, and health status. The evidence is also inconclusive about how WTP varies with cause of death, but there is greater clarity that reducing the risks of latent mortality response should be valued less than reducing risks of immediate death. **The committee concludes that the empirical evidence is insufficient to support a specific quantitative adjustment of WTP estimates to account for differences in remaining life expectancy, but it does not reject the general concept that such adjustments may be appropriate.** It is plausible that people with shorter remaining life expectancy would be willing to devote less of their resources to reducing their mortality risk than those with longer remaining life expectancy. In contrast, if the condition causing the shortened life expectancy can be treated and improved and an acceptable quality of life can be preserved or restored, people may put a high value on extending life, even if they have other health impairments or are quite elderly.

Recommendation: Researchers should continue to explore how WTP for reduction in mortality risk may vary with individual characteristics (such as age and health status), type of risk (such as accident vs illness), and time between changes in air quality and changes in risk.

Efforts to obtain better information about the preferences of the older population regarding reductions in mortality risk will probably entail greater use of surveys in which subjects respond to hypothetical situations.

Value of a Statistical Life, Individual Characteristics, and Risk Contexts

To estimate VSL, EPA mostly uses WTP values that are based on a context (for example, traffic accidents or workplace accidents) and a population that differ from the context and population relevant to the pollution-related risks that EPA is assessing. Two general approaches are available for measuring WTP for changes in mortality risk. The revealed-preference approach analyzes actual human behavior from which WTP for mortality-risk reductions may be estimated, for example, a wage-risk study of people's decisions on tradeoff between income and job-related mortality risk. The stated-preference approach surveys subjects' responses to hypothetical situations designed to reveal their WTP. **EPA's use of average WTP values in different risk contexts and for different population characteristics introduces considerable uncertainty about how these factors affect estimates of benefits. However, the current literature is inconclusive about how and how much the WTP values may vary with those factors. Although it is difficult to say how much the WTP values may differ, it is apparent that wage-risk studies cannot focus on the population and the risk context for ozone mortality.**

Recommendation: EPA should ensure that estimates of average WTP selected from the literature reflect results of both revealed-preference studies and stated-preference studies. EPA should consider the strengths and weaknesses of each study approach and consider how closely the available studies match the policy context in population at risk and type of risk. EPA should give less weight to wage-risk studies in selecting estimates of the WTP than in the past.

The Value of a Statistical Life Year

Given that the committee recommends the development of models for estimating life years saved (in addition to estimating changes in annual death rates and reductions in premature deaths), is it feasible to assign a monetary value directly to changes in life expectancy? Some analysts have converted VSL estimates into monetary values per statistical life year (VSLY) (see Chapter 2). Just as for VSL, there is a choice between using a constant or non-constant VLSY. Because a calculation of life years saved takes into account the remaining life expectancy of those whose deaths may have been prevented in a given period, use of a constant VSLY assumes that WTP values for mortality-risk reductions are consistently declining with increasing age. **Available empirical evidence does not support that assumption, so it does not support the use of a constant VSLY for benefits assessment. However, the economics literature does not reject the use of a non-constant VSLY (or a non-constant VSL). There is likely to be good reason to use non-constant values if there were sufficient empirical evidence of how WTP for mortality risk reduction varies with differences in remaining life expectancy and other factors. However, the committee concludes that current evidence is not sufficient to assign a non-constant VSLY or non-constant VSL.**

> **Recommendation: Unless future research produces empirical support for the assumptions that underlie a constant VSLY, EPA should not attempt to make valuation adjustments for changes in remaining life expectancy by estimating life years saved and applying a constant VSLY. More research is needed on appropriate ways to measure the values that people attach to changes in life expectancy.**

Sensitivity Analyses

Use of the average VSL obtained from the literature may overestimate the WTP to reduce ozone-related risk of premature death. That is because the population of older people appears to have greater mortality risk associated with ozone. Because older people have average remaining life expectancy that is substantially less than that of the whole population, the WTP to reduce the risk of death in the older population might be less than the WTP of the population as a whole. However, the effect of shorter life expectancy on older people's WTP may be offset to some extent by a higher WTP for a reduction in risk because of their poorer health status or their higher baseline risk compared with those of the general population. Results in the empirical literature are not consistent, but several studies suggest that WTP to reduce mortality risk is constant or declines slightly with age. **That implies that a proportional adjustment of the VSL for remaining life expectancy (that is, using a constant VSLY) would result in using too low a value of WTP for reducing ozone-related mortality.**

> **Recommendation: Given the uncertainty in the accuracy of available estimates of the VSL for ozone-related mortality, EPA should conduct sensitivity analyses that use a range of estimates or assumptions to see how the overall conclusions of the cost-benefit comparison might change. The selection of alternative assumptions for the sensitivity analyses could be based on either theory or evidence. However, when there is less confidence in the alternative assumptions used in sensitivity analyses, their results should not be given equal weight in the presentation of results.**

Recommendations for Future Research on Valuation

There is a fundamental need to understand better how age and remaining life expectancy affect WTP for reductions in mortality risk or increased life expectancy. An important next step is to ask that researchers report total age effects (WTP by age cohort) in addition to effects of age alone on WTP for a small reduction in mortality risk in a given period. Given the correlation of age with some of the other factors, there may be less uncertainty in the estimates of a total age effect. However, age-related income differences, sex differences, health differences, and the like would then be embedded in the estimates of WTP, and it might not be appropriate to use different VSLs that have these effects embedded.

Several recent economic studies have attempted to assess the effects of age and other factors on valuation. However, the efforts have been hampered by the lack of availability of the datasets produced for the published studies. EPA should urge researchers whom it funds for WTP studies to make their datasets available for future meta-analyses in addition to providing their published results.

EPA and the scientific community should explore and develop methods for characterizing and valuing changes in mortality risk that reflect the full life cycle. Studies to date have focused on WTP for annual changes in mortality risk, but the risk change of interest in most pollution-control assessments is more comprehensively described as a shift in survival probabilities across a large part of the human life span.

Environmental-benefits assessments rely primarily on estimates of WTP for reductions in risks of accidental death to estimate values for reducing risks of illness-related deaths. It is unknown how risk context (such as illness vs a work-related accident) affects a valuation estimate. EPA and the scientific community should seek to learn more about how mortality-risk characteristics affect the valuation of reducing risks.

FUTURE REGULATORY-IMPACT ANALYSES INVOLVING OZONE MORTALITY

Because short-term exposure to ambient ozone is likely to contribute to premature deaths, future regulatory-impact analyses (RIAs) concerning ozone-control measures should include the benefits of reduced mortality risk. As in EPA's RIA for the recently finalized ozone NAAQS, emphasis should be on using estimates from new systematic multicity analyses that used national databases of air pollution and mortality, such as in the National Morbidity, Mortality, and Air Pollution Study, without excluding consideration of meta-analysis of previously published studies. Future RIAs should give little or no weight to the assumption that there is no causal association between estimated reductions in premature mortality and reduced ozone exposure. Health-benefits estimates should be accompanied by a broad array of analyses of uncertainty.

Distributed-lag models over several days appear better than single-day models at capturing the acute and subacute mortality effects of ozone exposure and should be part of future benefits assessments to the extent that they are supported in the literature.

Future RIAs should incorporate research results on the mortality effects of chronic ozone exposure and research that addresses key uncertainties related to potential confounding factors, exposure measures, and susceptibility as appropriate.

Despite many concerns about the accuracy of any specific WTP value and a corresponding VSL that does not vary with population or risk characteristics, the committee recommends a single VSL as the most scientifically supportable approach at present for monetary valuation of ozone-related mortality. Before making a substantial change in its approach for valuation of mortality-risk reductions, EPA should have fairly conclusive empirical evidence to support the change. It is the committee's judgment that the available evidence is not now sufficient to support such a change, but sensitivity analyses should explore alternative approaches and further research should be conducted to answer the questions raised about the validity of EPA's current approach. Benefits-assessment methods may need to be revised as new information emerges on characteristics of populations susceptible to an ozone-mortality risk and on variations in WTP for mortality-risk reductions (or increases in life expectancy) based on different population characteristics.

EPA should consider placing greater emphasis on reporting changes in age-specific death rates and changes in life expectancy in the relevant populations than on reporting estimates of lives saved or premature deaths avoided.

In this report, the committee has identified major gaps in knowledge about methods for assessing benefits of ozone-related mortality risk reduction and has recommended research strategies to close those gaps. The committee recognizes that many of the recommended research activities are complex and will be difficult to undertake, and that sufficient resources may not be available to undertake them all in the near term. Therefore, EPA and other agencies that might carry out the recommended research will need to set priorities and develop a strategy for addressing the various information needs.

1

Introduction

The Clean Air Act requires the U.S. Environmental Protection Agency (EPA) to issue and review periodically the National Ambient Air Quality Standards (NAAQS) for each of six criteria pollutants. One of those pollutants is ozone in the lower atmosphere.[1] Although ozone was originally viewed as an urban pollutant (associated largely with such locations as Los Angeles), it is now recognized that ozone formation and transport over much larger areas result in increased exposures to humans nationwide (e.g., NRC 1991). International transport of ozone exacerbates the problem (Jacob et al. 1999).

Efforts to mitigate ambient ozone involve controlling precursor emissions (nitrogen oxides [NO_x] and volatile organic compounds [VOCs]) from a wide range of stationary sources (such as factories, electricity-generating facilities, and gasoline stations) and mobile sources. Although EPA is not allowed to consider costs and benefits when setting NAAQS, economic analyses are carried out for the proposed and final NAAQS to estimate costs and benefits expected to result from the standards. And federal agencies deciding on (generally national) actions expected to cost more than $100 million per year are required to carry out a cost-benefit analysis of alternative regulatory strategies and to assess the distribution of their impacts in different segments of society.

In past assessments of the benefits of regulations to mitigate ozone, EPA has addressed the relationship between ozone exposure and respiratory disorders. EPA's reviews since the Regulatory Impact Analysis for the 1997 NAAQS (EPA 1997a) have included the results of epidemiologic studies that link ambient ozone concentrations with premature deaths in sensitivity analyses, but EPA has not included these mortality results in their primary estimates of the benefits

---

[1]In this report *ozone* is used to refer to the broad array of photochemical oxidants present in ambient air.

of reductions in ambient ozone concentrations because of uncertainty about the appropriate interpretation of these results.

Studies published since 1990 have yielded evidence in support of a relationship between short-term exposure and premature death. They include city-specific time-series studies, many of which have been included in recent meta-analyses (Bell et al. 2004, 2005; Ito et al. 2005; Levy et al. 2005). However, interpretation of the evidence is complicated by ozone's occurrence in mixtures of other pollutants whose concentrations fluctuate in a similar manner and can be influenced by short-term meteorologic changes. It is important to understand whether it is ozone or a copollutant that is causing health effects, because steps to control the wrong agent will be both expensive and ineffective. Interpretation of the health studies results is also complicated by uncertainties that result from relying on ozone measurements from outdoor monitoring sites to estimate exposures of people who spend most of their time indoors.

If changes in risks of premature mortality are attributed to changes in ambient ozone concentrations in regulatory benefits assessments, EPA also needs an estimate of the monetary value of such changes in mortality risk. In previous sensitivity analyses of mortality risk reduction benefits from ozone reductions, EPA used the same value of statistical life (VSL) as that used for analysis of other mortality risks, such as those associated with particulate matter with a diameter less than or equal to 2.5 microns ($PM_{2.5}$). Available VSL estimates are drawn largely from studies of working-age adults with average remaining life expectancies. There are many questions about the applicability of these estimates to mortality risks associated with ozone, which may fall disproportionately to an older population with shorter remaining life expectancy and more frail health status.

CHARGE TO THE COMMITTEE

In light of the recent evidence on ozone mortality risk and questions about its implications for benefit analysis, EPA asked the National Research Council for scientific advice on how the ozone-mortality research findings could be used in the context of health-benefit analyses associated with regulatory assessments. In response, the National Research Council established the Committee on Mortality Risk Reduction Benefits from Decreasing Tropospheric Ozone Exposure (see Appendix A).

The Statement of Task to the committee (see Box 1-1) includes evaluation of the scientific and technical bases of approaches used by EPA for estimating reductions in mortality risk and associated benefits of health-based ozone standards over time. The committee was to assess methods for estimating reductions in premature death due to diminished short-term exposure to ozone, increases in life expectancy, economic valuation of the increased life expectancy, and associated uncertainties and their general implications for decision-making. The

BOX 1-1 Statement of Task

An NRC committee will evaluate the scientific and technical bases of approaches used by EPA for estimating ozone mortality risk reduction and associated benefits of health-based standards over time. It will assess methods for estimating reductions in premature mortality due to diminished short-term exposure to ozone, increases in life expectancy, economic valuation of the increased life expectancy, and associated uncertainties and their general implications for decision making. In addition, the committee will recommend approaches for characterizing and communicating each of those aspects in regulatory health benefit analyses.

Specifically, the committee's evaluation will include consideration of the following aspects:

The committee will take into account the relevant reports of past NRC and IOM committees, as well as risk assessment for particulate matter (PM) carried out by the EPA.

(1) Relative contributions of recent studies (especially those published since 2003) for characterizing the size of the ozone-mortality effect in the context of benefits analysis.

(2) Potential implications of methods used in the recent studies on reported benefits estimates (e.g., selection of data considered in the studies, selection of mortality effect estimates from within the considered data, control for effect moderators [such as temporal trends], treatment of potential publication bias, or other factors). Include consideration of the likely direction and magnitude of any influences that may result from choice of methods.

(3) Available data and methods to account for the relative roles of potential confounders, such as copollutants, and the influence they have on estimates of ozone mortality effects.

(4) The most appropriate exposure metrics for use in developing a concentration-response function designed to quantify the number of premature mortalities associated with specific regulatory options. How does the choice of exposure metrics affect benefits estimation? Include consideration of dose estimation, particularly for sensitive groups.

(5) Adequacy of a basis for estimating the likely impact on life expectancy from reductions in short-term daily exposures to ozone. If there is an adequate basis, provide specific recommendations, if possible, on approaches for how EPA should quantitatively express the magnitude and associated uncertainties of this impact.

(6) Strengths and weaknesses of the data, methods, and assumptions employed by EPA for sensitivity analyses in the context of an ozone-mortality effect relationship.

(7) Quantitative approaches for estimating the degree of uncertainty in premature mortality estimates associated with reductions in ambient

(Continued)

BOX 1-1 Continued

ozone concentrations. Aspects to be considered will include both parameter and model uncertainty in the concentration-response function, differences within and between the epidemiologic and toxicologic data bases, reliance on surrogates of personal exposure, identification of sensitive populations, and other factors. Identify methods of presenting the quantitative uncertainty estimates.

(8) Scientific approaches for assigning economic values to reductions in mortality risk when the reduction in risk results in changes in life-expectancy of varying lengths. If there is an adequate basis for quantifying those changes, consider:

(a) How the current understanding of premature mortality associated with short-term ozone exposure informs the most appropriate scientific approach to economic valuation of benefits. Consider methods to account for baseline health status, baseline and the magnitude of the change in life-expectancy, differences across various sectors of the population, uncertainty, and other relevant factors.

(b) Applicability of the economics literature on mortality risk valuation and the associated methodologies developed for economic valuation of premature mortality reductions in the context of the health effects of ozone.

(c) Issues that are specific to ozone and those which may be more broadly applicable to benefits assessments for other risk reduction strategies.

(9) Major gaps in knowledge about ozone-mortality benefits analysis and the most promising research strategies to close those gaps. Identify any additional data, analyses, or research needed to separate the relative contributions of ozone and other gaseous or particulate components of the air pollution mix to the total short term premature mortality effect documented in the literature.

In assessing the methods for estimating ozone mortality risk, the committee will not develop its own estimate of such risk. In assessing the methods for economic valuation of reductions in mortality risk, the committee will not itself judge the appropriateness of values assigned to a life saved or life years added.

Sponsor: U.S. Environmental Protection Agency.

committee was asked in the Statement of Task to consider the relative contributions of recent studies (especially those published since 2003) to characterization of the size of the ozone-mortality effect in the context of benefit analysis. The recent studies include Bell et al. (2004); Bell et al. (2005); Ito et al. (2005); and Levy et al. (2005). The committee was also to consider the potential implications of the methods (such as criteria for selection of data) used in the recent studies.

The committee was asked specifically not to develop its own estimate of ozone mortality risk or to judge the appropriateness of values assigned to a life saved or life years added by decreases in ozone exposure. Because the committee was asked to focus on human mortality risk reduction and associated benefits, it considered changes in human morbidity as an end point only to the extent that they illuminate issues related to mortality. The committee was not asked to assess changes in other effects of ozone (referred to as secondary effects), such as effects on agricultural crops and forests that would result from changes in ozone concentrations.

In carrying out its charge, the committee considered presentations made during its public sessions and relevant scientific and technical documents prepared by EPA and other organizations. The committee sought to benefit from and build on the work of other advisory groups. For example, the National Research Council report titled *Estimating the Public Health Benefits of Proposed Air Pollution Regulations* evaluated methods used by EPA to estimate health benefits, primarily the analysis of mortality associated with exposure to airborne particulate matter (NRC 2002). The Institute of Medicine report *Valuing Health for Regulatory Cost-effectiveness Analysis* (IOM 2006) addressed analytic and policy issues, including the use of cost-effectiveness analysis to estimate health-related effects of regulatory actions. EPA's Science Advisory Board (SAB) has completed several reviews of ozone mortality in EPA's benefit-analysis framework. An SAB developed an advisory report on mortality risk valuation and how the valuation may be affected by differences in life expectancy (EPA SAB 2007). The journal *Review of Environmental Economics and Policy* published an issue containing a series of articles on mortality risk and valuation (Review of Environmental Economics and Policy 2007, Vol. 1[2]). The report of a conference on Critical Considerations in Evaluating Scientific Evidence of Health Effects of Ambient Ozone summarized scientific issues the conference participants viewed as central for decisions to be made in setting the ozone NAAQS (Brauer et al. 2007). The committee also considered a EPA staff paper that summarized staff recommendations on the ozone NAAQS to the EPA administrator (EPA 2007a) and EPA's regulatory impact analysis to accompany the agency's recent proposal for revising the NAAQS for ozone (EPA 2007b).

Figure 1-1 presents a sequence of analytic methods used by EPA to assess mortality risk reduction and associated health benefits expected to result from setting and implementing ozone NAAQS (see Chapter 2). Estimation of the potential health effects of actions taken to improve air quality, and the economic value of those effects, requires estimation of the likely changes in emissions, ambient ozone concentrations, population exposures, and health risk and then the economic valuation of the change in risk. Figure 1-1 shows which components of the committee's Statement of Task correspond to specific steps in the sequence of analytic methods. It also indicates which chapters of this report cover specific topics. Chapter 6 integrates the committee's overall conclusions and recommendations concerning ozone exposure, mortality risk, and benefit assessment.

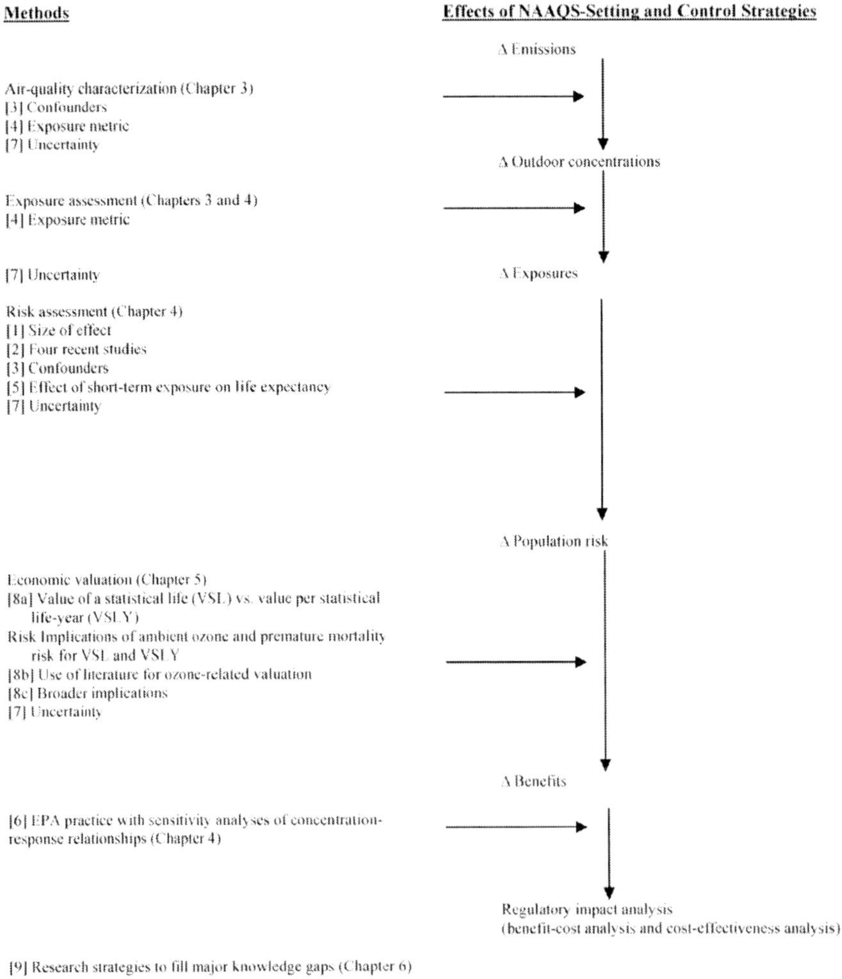

FIGURE 1-1 Methods used by EPA to assess the effects of NAAQS and control strategies developed to implement the standards. The left portion of the figure indicates the chapters of this report that deal with each method. Numbers in brackets correspond to relevant portions of the committee's statement of task, shown in Box 1-1. NOTE: The committee was not asked to and did not consider the cost aspects of benefit-cost analysis and cost-effectiveness analysis. Abbreviations: Δ = change; NAAQS = National Ambient Air Quality Standard.

2

Overview of Ambient-Ozone Standards Development and Benefits Assessment

INTRODUCTION

Ozone has been the subject of extensive research, standard-setting, and air-pollution control activities for over 3 decades in the United States and in other countries. This chapter provides background on the committee's consideration of the implications of recent findings of associations of ambient ozone concentrations with mortality, including the setting of national ambient air quality standards, the process of implementing the standards, the scientific basis of standard-setting, the use of health studies to quantify expected health effects of changes in air quality, the concepts underlying economic valuation of such health effects, and experience in the United States and elsewhere in applying these concepts to estimate benefits of ozone reduction.

SETTING NATIONAL AMBIENT AIR QUALITY STANDARDS FOR OZONE

Beginning in 1970, the U.S. Clean Air Act (CAA) directed the Environmental Protection Agency (EPA) to consider the best available science bearing on exposure to and effects of several ambient air pollutants that are emitted by a wide array of sources and to set National Ambient Air Quality Standards (NAAQS) for pollutants to which the public was widely exposed. Under the CAA, NAAQS are to be set and reconsidered every 5 y, and the administrator of EPA is to consider setting *primary* NAAQS to protect the public health and *secondary* NAAQS to protect the public welfare (for example, buildings, materials, and ecosystems). In setting the primary NAAQS, the administrator is to consider all available science, historically compiled in a so-called criteria document[1] prepared by EPA scientists and consultants, and to set standards at levels that are "requisite to protect the public health with an adequate margin of safety." The act does not include costs of implementation in the criteria for setting the NAAQS, and EPA—with consistent support from the courts—has interpreted that to mean that it may not consider costs in NAAQS decisions (American Trucking Associations, Inc. v. U.S. EPA, D.C. Cir 97-1440 and 97-144).

In essence, that has meant that each time a NAAQS is reviewed, the administrator must weigh the most recent evidence and continuing uncertainties and make a "public-health policy judgment" about whether the newest evidence provides enough certainty about the likelihood and public-health significance of effects above, at, and below the current standard to warrant a determination that the current standard is adequate to protect the public health with an adequate margin of safety or should be lowered or raised.

Once that determination is made for a particular pollutant or class of pollutants, EPA is expected to make decisions about four aspects of the standards:

- The ***indicator*** (the pollutant to be monitored and assessed for attainment).
- The ***level*** of the standard.
- The ***averaging time*** (for example, 1 h, 8 h, 1 d, or 1 y).
- The ***statistical form of the standard*** (for example, whether the standard will not be met if it is exceeded more than 1% of the time, on the third-highest day each year, or other similar measure).

Since the inception of NAAQS, EPA has determined that photochemical-oxidant air pollution, formed when specific chemicals in the air react with light and heat, is of sufficient public-health concern to merit establishment of a primary NAAQS. In implementing that determination, EPA has since 1979 identified ozone, a prominent member of the class of photochemical oxidants (such as nitrogen dioxide), as an indicator for setting the NAAQS and tracking whether areas of the country are in attainment of the standards. Over the last 38 y, as scientific understanding of ozone health effects has evolved, EPA has reviewed and updated, as needed, the primary NAAQS for ozone and other photochemical oxidants in 1971, 1979, 1993, and 1997 (see Table 2-1). As illustrated in Figure 2-1, under the 1997 NAAQS, ozone nonattainment has occurred largely in heavily populated areas east of the Mississippi River,

---

[1]In December 2006, EPA indicated that after the current ozone NAAQS review process, it would no longer use the historical terminology of *criteria document* to summarize the science and *staff paper* to summarize staff risk assessment and recommendations to the administrator. The criteria document would be replaced by an *integrated science assessment* and the staff paper would be replaced by an *advanced notice of proposed rulemaking* (EPA 2007c).

TABLE 2-1 History of Primary NAAQS for Ozone and Other Photochemical Oxidants[a]

Year	Indicator	Level	Averaging Time	Form	Scientific Rationale
1971	Total photochemical oxidants	0.08 ppm	1 h	Not to be exceeded more than 1 h/y	Principal study cited for final found increased asthma-attack frequency when hourly average reached 125μg/m^3 (0.10 ppm); final standard included margin of safety below most likely threshold suggested by this study (Bachmann 2007)
1979	Ozone	0.12 ppm	1 h	More than 1 d/y with maximal hourly average above 0.12 ppm	Move to ozone-specific effects; three clinical studies found reduction in pulmonary function or symptoms as having lowest effect level in humans at 0.15-0.3 ppm with evidence of lower effect levels in animals (Bachmann 2007)
1993	Ozone	Determination that no change was necessary			
1997	Ozone	0.08 ppm	8 h	Annual 4th-highest daily maximal 8-h average concentration, averaged over 3 y	Many new studies over a decade found effects at concentrations below 1979 standard with increased importance of 6-8-h exposures; key studies found lung-function decrements, respiratory symptoms, increased sensitivity to irritants, indicators of pulmonary inflammation increasing across range of 0.08, 0.1, and 0.12 pm for 6- to 8-h exposures with subjects engaged in intermittent exercise; numerous epidemiologic studies found increased hospital admissions and emergency-room visits for respiratory causes attributed primarily to ozone; proposal highlighted risk assessment for children (nine-city) and for New York City hospital admissions. (Bachmann 2007)
2008	Ozone	0.075 ppm	8 h	Annual 4th-highest daily maximal 8-h average concentration, averaged over 3 y	Strengthened clinical database showing effects at 0.080 ppm; limited data on potential effects on some subgroups at 0.060 ppm (6- to 8-h exposure) (Adams 2002, 2003, 2006); substantially increased epidemiologic evidence on morbidity effects (EPA 2008a). Also see Box 2-1.

[a]This table focuses on the primary rather than secondary NAAQS.

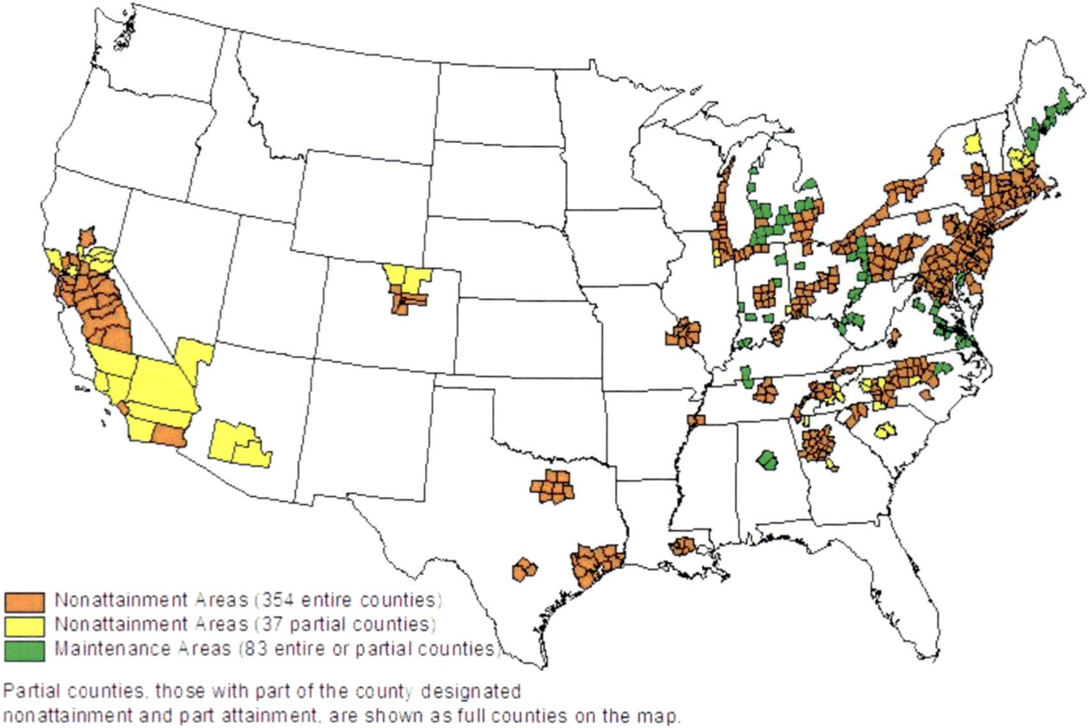

FIGURE 2-1 Counties violating 1997 primary 8-h NAAQS for ozone and other photochemical oxidants. Source: EPA 2007d.

BOX 2-1 The Estimation of the "Policy Relevant Background"

One step in the process of setting the NAAQS which was not a focus of this report is to assess the amount of risk reduction expected to result from its implementation, an involved and often ambiguous process. It is unlikely that lowering a standard will lead to no ozone concentrations occurring that are greater than the set level without ozone concentrations below the level of the standard also being affected. Likewise, raising the level of the standard would not lead quickly to ozone concentrations increasing up to the standard at all locations. In order to conduct its risk assessment, EPA assesses how ozone at all concentrations will respond to the change in the standard. However, during the NAAQS process the specific rules and controls that will be implemented in response to a change are not known, so neither is it known how a change in the standard will impact air quality. There are a large number of possible paths towards cleaner air and attaining a standard, and as discussed later, their impacts on ozone concentrations at different times of the day and different periods of the year can differ substantially, even if all of the paths suggest that they will lead to attainment. The question is how can EPA assess the health and welfare implications of a revised NAAQS without better knowing the likely air quality response.

EPA deals with this ambiguity by prescribing a "Policy Relevant Background" (PRB), and then rolls back current levels towards this background level in such a way that the standard would now be met. In so doing, all levels above the PRB are reduced (see EPA [2006a, 2007a] for a more complete description of the definition, calculation and use of PRB). EPA further goes on to define the PRB as the level that the pollutant concentration would be in the absence of anthropogenic emissions from the US, Mexico and Canada. In the 2006 Ozone Criteria Document, a global chemical model was used to calculate spatially and temporally varying PRBs (EPA 2006a). Removing emissions from Canada and Mexico has been criticized for, among other reasons, providing an unrealistically low estimate for PRB (Brauer et al. 2007).

Defining a PRB is unnecessary for establishing a level above which ozone is harmful to human health or for estimating changes due to a specific regulatory action, nor is it needed for quantifying human health responses to short-term ozone levels or monetizing risk changes in response to ozone changes, which are the primary foci of this study. Thus, this Committee did not explore this issue in detail, though the report discusses aspects of PRB in Chapter 3.

throughout California, and in major cities in Texas, Arizona, and Colorado. Figure 2-2 shows counties with monitor readings that would violate alternate 8-h ozone standards of 0.070 and 0.075 ppm proposed by EPA in June 2007.

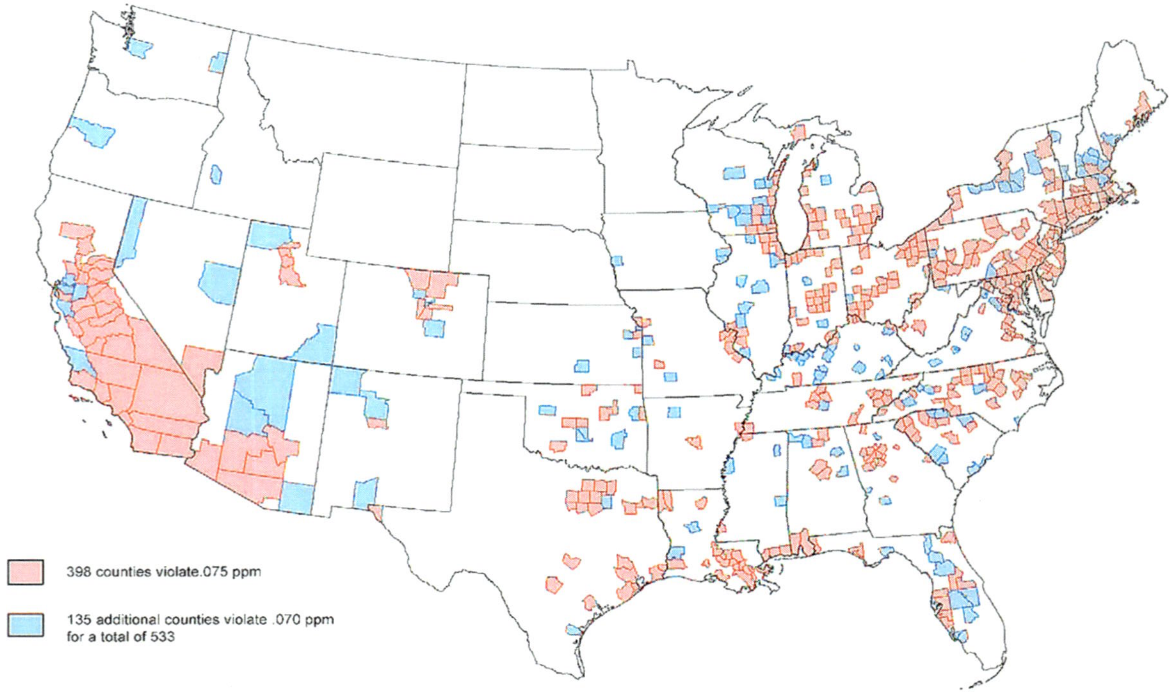

FIGURE 2-2 Counties with monitors readings that would violate alternate 8-h ozone standards of 0.070 and 0.075 ppm proposed by EPA in June 2007 (on basis of 2003-2005 monitoring data). Source: EPA 2007e.

On March 12, 2008, EPA issued revisions to the ozone primary and secondary NAAQS (EPA 2008a). The level of the primary 8-h standard was lowered to 0.075 ppm and the secondary standard was set to be the same as the revised primary standard.

IMPLEMENTING THE NAAQS

Once a NAAQS is set, EPA and the states and tribes pursue three basic paths to ensure that the standards are met throughout the country:

- First, EPA establishes reference methods for measuring the indicator pollutant, and states and tribes implement monitoring programs to determine whether areas in their jurisdictions are in or out of attainment of the NAAQS. Under the 1997 ozone NAAQS, about 391 counties or parts of counties with a total population of over 36 million are in nonattainment of the standards (Figure 2-1).
- Second, once EPA determines that any area is not in attainment of the NAAQS, the state that contains the area must develop a state implementation plan (SIP) that includes the actions it will take to reduce emissions and bring the area into attainment. States that are in attainment must consider actions that will ensure that they maintain that status. If a state fails in those efforts, EPA has the authority to implement its own plan; this has rarely occurred.
- Third, to the extent that Congress has authorized EPA to take national or regional action to address emissions that lead to nonattainment, EPA proposes and implements regulations to do so (for example, national motor-vehicle and fuel standards).

Beyond those regulatory actions, EPA works with states, tribes, and local authorities to implement an air-quality index system, which assesses the likelihood that air quality in a given area on a given day will be near or above the standard. The likelihood of poor air quality in turn sets off broad-based public-information efforts to alert residents, especially such sensitive populations as the elderly or asthmatic, to restrict activity that might increase exposure.

In the case of ozone, regulatory actions at state and federal levels have taken many forms and have resulted in controls on emissions of volatile organic compounds and nitrogen oxides—the two primary precursors of ozone and other photochemical oxidants—from fuels, on-road and nonroad engines, electric-power facilities, manufacturing facilities, consumer products (such as paints), and many other sources. A National Research Council analysis of those efforts in 2004 found that although many actions have been taken at federal, regional, and state levels, the vast majority of reductions have come through national and regional multistate actions, rather than individual state actions. An exception is California, which has had the most aggressive air-quality programs in the nation (NRC 2004a).

THE SCIENTIFIC BASIS OF PRIMARY NAAQS FOR OZONE

For over 50 y, detailed scientific investigations have examined nearly every aspect of ozone and other photochemical oxidants, and this robust body of research has served as the basis of each succeeding NAAQS determination by EPA. The research has taken several forms:

- *Atmospheric chemistry* to understand the sources of ozone precursors and the formation of ozone and other photochemical oxidants.
- *Exposure research* to characterize pollutant concentrations and human activity patterns that lead to contact with the pollutants over specific periods. *Toxicologic research* in vitro test systems (such as various human cells and other biologic media) and in vivo experiments (involving a wide array of animal species and animal models of human disease).
- *Epidemiologic research,* including panel, cohort, and broader population studies of the acute or chronic effects of ozone exposure on morbidity and mortality in children and adults.
- *Controlled human exposure (or "clinical") studies* of varied healthy and compromised adult volunteers exposed to ozone or ozone combined with other gases.

Research findings have contributed to the understanding necessary to establish a NAAQS. In particular, the human studies—epidemiologic and clinical studies—have contributed several key pieces of standard-setting information, including information on the effects of real-world exposure of human beings to contaminants, the smallest exposure that can be demonstrated to have effects in humans, and the degree to which effects are found in sensitive populations.

Recent Scientific Conclusions About Risks Posed by Ozone Exposure

Overall, extensive toxicologic, epidemiologic, and clinical research on the effects of exposure to ozone has yielded strong evidence of effects on respiratory end points at or near current regulatory levels, and in a variety of populations. However, clinical studies have shown substantial variability in individual responses; even sensitive populations, such as asthmatics, exhibit a range of responses and nonresponse (EPA 2006a). Although an extensive literature—in toxicology, epidemiology, and controlled human exposure studies—has reported consistent evidence of such effects, there are still uncertainties in our understanding, for example:

- Uncertainty about the relationship between outdoor ozone concentrations and personal exposure, especially of persons who spend most of their time indoors or use air conditioning during periods of peak ozone concentrations.

• Uncertainty about the degree to which epidemiologic studies can identify the effects of individual pollutants, such as ozone, that are present in mixtures of pollutants in the ambient air.

• Uncertainty about the degree to which current study designs and datasets enable understanding of risks resulting from exposures at very low pollutant concentrations and whether there is any threshold for effects.

• Uncertainty, despite high-quality studies of long-term effects on lung growth in children (CalEPA 2006), about long-term exposure to ozone as a risk factor for chronic illness and premature death.

On the basis of the available evidence, EPA in its most recent criteria document (EPA 2006a, p. 8-77) concluded that "the overall evidence supports a causal relationship between acute ambient ozone exposures and increased respiratory morbidity outcomes resulting in increased ED (emergency-department) visits and respiratory hospitalizations during the warm season." EPA's most recent staff paper (EPA 2007a, pp. 5-92 and 5-93), summarizing staff recommendations on the NAAQS for the administrator) further concludes that there is clear and convincing evidence of causality for lung function decrements in healthy children under moderate exertion for 8-hr average ozone exposures. We also judge that there is strong evidence for a causal relationship between respiratory symptoms in asthmatic children and ozone exposures and between hospital admissions for respiratory causes and ambient ozone exposures. There is greater uncertainty and somewhat less confidence about the relationship between ozone and non-accidental and cardiorespiratory mortality, although the Criteria Document's overall evaluation is that it is highly suggestive that this relationship exists. The strengths and weaknesses of the recent literature on mortality and ozone are explored in greater detail in Chapter 4.

Recent reviews of the evidence by the World Health Organization in its establishment of world air-quality guidelines (WHO 2006) and by the California Air Resources Board in its establishment of state air-quality standards (CalEPA 2007) have reached similar conclusions although the latter reviews have given somewhat greater causal weight to the associations with premature mortality.

Evolution of the Science of the Health Effects of Ozone and Particulate Matter

A broad and deep literature on the health effects of exposure to ozone has developed over the last 35 y and has resulted in a series of actions to set and revise the NAAQS and to reduce emissions of precursors. Similarly, over the last 15 y, much scientific and regulatory activity has been generated by intense interest in the effects of exposure to particulate matter (PM), especially PM with a diameter equal to or less than 2.5 µm ($PM_{2.5}$) (EPA 2007f). There are useful parallels and differences between the literature on ozone and that on $PM_{2.5}$, and they have to some extent resulted in different approaches to the estimation of benefits of reducing exposure and in differences in actions taken.

In general terms, the ozone literature developed first out of observations of populations exposed outdoors (for example, panel studies of summer campers) and then from what has become a large set of controlled human exposure (clinical) studies. More recently, and to some extent in parallel with the PM literature, ozone research has moved into the realm of broader population epidemiology with a growing number of hospitalization studies and other time-series studies starting in the 1990s. Cohort studies have not been designed to look specifically at ozone and premature mortality, and testing for the statistical significance of ozone-mortality relationships was not feasible due to insufficient variation in estimated long-term exposure to ozone (see Chapter 4). However several studies, especially the Southern California Children's study, have examined longer-term morbidity effects (Tager et al. 2005; CalEPA 2006). Most cohort studies of PM have examined potential effects of ozone both as a confounder and for its independent effects, although few have found statistically significant evidence of such effects (see Chapter 4 for a review of the studies). One regulatory result of the relative absence of data on longer-term effects has been a focus on setting only a short-term NAAQS for ozone: a 1-h standard, which after nearly 30 y was replaced by an 8-h standard in 1997.

The $PM_{2.5}$ literature expanded rapidly after findings in the early 1990s of the associations of PM with morbidity and mortality in single-city time-series studies followed by the publication of results of two major cohort studies—the Harvard Six Cities Study (Dockery et al. 1993) and an American Cancer Society (ACS) study (Pope et al. 1995)—that showed much larger associations with longer-term residence in areas with high concentrations of $PM_{2.5}$. Those results motivated the first major revision of the PM NAAQS in a decade in 1997 with the development of both shorter-term (24-h) and longer-term (annual) standards and a host of aggressive new regulatory actions for vehicles, power plants, and other sources. The controversy over those standards resulted in the launching of a major multidisciplinary research program (over $50 million/y for nearly a decade), which has provided substantial additional information on exposure, toxicology, and epidemiology and involved the first use of clinical studies in this context (NRC 2004b). In turn, the new information led to many specific control actions related to such sources as diesel engines (EPA 2007g) and electric-power plants (EPA 2007h) and, in September 2006, an EPA decision to tighten the daily PM standard (but not the annual standard). Those actions were justified largely by cost-benefit analyses based on the substantial potential mortality benefits estimated from the ACS study and later reanalyses and extended analyses (Krewski et al. 2000; Pope et al. 2002, 2004).

The focus on PM science and regulation over the last decade has had several ramifications for ozone. The understandably enhanced policy attention to PM resulted in substantially increased funding for and scientific attention to PM research, to some extent at the expense of research on ozone and other pollutants (except as they might be confounders of PM effects). The substantial estimated mortality benefits of reducing PM have resulted in estimates of health benefits of reducing PM that are much larger than those estimated for reducing ozone and other pollutants; one consequence has been in substantial political support for revisions in regulations that address PM. In the last several years, partly because of the increased attention to mortality, there have been increased efforts on the part of regulators and scientists to understand better whether mortality effects may be attributed to exposure to ozone and, if so, to incorporate them into future standard-setting and benefits assessment. Those considerations caused EPA to fund of studies of the short-term mortality effects of ozone that are a primary focus of the present committee's review.

REGULATORY BENEFITS ASSESSMENT FOR SETTING AND IMPLEMENTING NATIONAL AMBIENT AIR QUALITY STANDARDS

Quantitative assessment of the potential health benefits of actions planned to improve air quality has become a central component of regulatory impact analysis in the United States and other countries. As described in Figure 1-1 (see Chapter 1), the process involves estimation of the likely changes in population exposure to pollutants and the resulting changes in health risk as well as the economic valuation of the changes in health risk.

Use of Health Studies for Regulatory Benefits Assessments

There is a fundamental difference between information needed for regulatory benefits assessment and information needed to set a protective health standard. Selection of a primary (health-based) standard focuses on the lowest ambient concentration that poses a risk of adverse health effects in the most sensitive population. Assessing benefits requires information for estimating all the reductions in health risks in the entire population that is expected to experience a reduction in ambient concentrations.

The information needs for a comprehensive benefits assessment have led EPA to rely almost entirely on epidemiologic studies to support quantification of health effects of PM and ozone, whereas clinical and toxicologic studies are often important for standard-setting. Epidemiologic studies link health outcomes measured in the general population or in specific cohorts to differences in ambient pollution concentrations at different times or in different locations. Widespread monitoring has made large-scale epidemiologic studies of the criteria pollutants feasible in the United States and many other countries. The real world thus becomes the laboratory, and subjects are studied in their normal environments. That has three important advantages for benefits assessment:

- Estimates of changes in rates of health outcomes in the population as a result of changes in ambient pollutant concentrations can be directly estimated without the extrapolations required when clinical or toxicologic study results are used for such purposes.

- Epidemiologic studies generally include serious health effects, such as death, chronic disease, and hospitalization that cannot be tested in clinical studies.
- Epidemiologic studies can be designed to measure the effects of long-term exposures to air pollutants, which are difficult to assess in clinical studies.
- At the same time, important limitations of epidemiology studies also exist, and these contribute to some of the key uncertainties in the benefits assessments for PM and ozone that EPA conducts for NAAQS and other rulemakings.
- Epidemiologic studies can demonstrate statistical associations, but alone they cannot prove causation. A number of factors make such casual determinations difficult, including potential confounding factors such as weather, other pollutants, and socioeconomic status; and uncertainties in assessing actual personal exposure. These factors can and have been included in analyses, but there is always a chance that their effects have not been adequately defined and removed.
- It is difficult to isolate the independent health effects of individual pollutants that occur in mixtures in ambient air. That is a concern for ozone and $PM_{2.5}$, especially in the warm-weather months in the eastern United States.

Those strengths and weaknesses of the epidemiologic literature, especially in the case of the ozone-mortality studies that are the subject of this report, are described in further detail in Chapter 4.

Concepts of Economic Valuation of Mortality Risk Reduction for Regulatory Benefits Assessment

A charge to the present committee is to address the quantification and economic valuation of mortality-risk reductions in the context of cost-benefit analysis. *Cost-benefit analysis* is the comparison of a monetary measure of the welfare gain for those who benefit from a policy or program with a monetary measure of the welfare loss for those who are harmed (that is, who incur costs) by the policy or program. Those who benefit and those who are harmed may be the same or different members of the population. For example, if a regulation causes higher electricity prices by reducing power-plant emissions, many of the people who pay higher prices will also experience a reduction in health risks because air quality is improved. However, the additional costs and the value of the health-risk reduction will not necessarily be the same for everyone affected. In general, cost-benefit analysis focuses on aggregate costs and benefits, and a program is considered to have a net benefit to society if total benefits exceed total costs. If an assessment of how costs and benefits are distributed is done, it is a separate assessment.

As noted above, cost-benefit analysis is intended to measure how those affected by a policy value the costs that they incur or the benefits that they receive. The appropriate monetary measure of a benefit is the maximal expenditure (or income reduction) that one would be willing to make to obtain the benefit if such a transaction were feasible. That is called the willingness to pay (WTP) for the benefit. The WTP in this context is a metric for the opportunity-cost value of the benefit because it is a measure of the alternative consumption opportunities that a person is willing to forgo to obtain the benefit.

WTP measures are based on studies of people's own preferences, including their preferences regarding tradeoffs between reducing risks to their own lives and other uses of their available resources. WTP values are not independent of the valuation context or of a person's circumstances, so a WTP value for a given benefit may vary among contexts and among people. For example, a WTP value is a function of a person's available income and wealth. That is inherent in the measure: it is a measure of the resources that a person is willing to exchange for a given benefit, and it will therefore be a function of the total resources available to that person. What that means for policy analysis is that WTP estimates should be drawn from studies of populations that have relevant characteristics similar to those of the population affected by the policy.

The cost-benefit analysis of a program that provides reductions in mortality risks typically focuses on the values of a risk reduction to all the affected persons summed to obtain the total benefit of the risk reduction. Questions are sometimes raised about whether values that people hold for others' risk reductions should be added to obtain the total value of a program to society. Jones-Lee (1991) presented a theoretical analysis that clarified that issue and illustrated the importance of unambiguously defining whose risk is being reduced in an empirical WTP estimation study. His analysis demonstrated under what conditions WTP for other people's mortality-risk changes should be added to the value of one's own risk reduction to obtain an optimal allocation of resources to health and safety efforts. His analysis showed that if people's altruism toward others is general regarding the others' enjoyment and satisfaction, there is no need to add WTP values for others' mortality risk reductions to WTP for people's own mortality risk reduction in a cost-benefit analysis. It is only in the case of what Jones-Lee calls "paternalistic altruism" that an argument could be made for adding WTP values that a person has for another's mortality risk reduction in a cost-benefit analysis. What he means by paternalistic altruism is when a person values the other person's consumption or behavior for its own sake regardless of the other person's enjoyment, such as when a parent is happy that the child eats the spinach even if the child is miserable doing it.

Estimates of the *value of a statistical life* (VSL) are often used in cost-benefit analyses of programs that are expected to reduce mortality risk in a population. VSL estimates are derived from estimates of a person's WTP for changing his or her mortality risk in a given period by a small amount (see Box 2-2). Although the VSL estimate is applied in an assessment by estimating the

BOX 2-2 Definition of Value of a Statistical Life in Relation to Willingness to Pay, and a Numerical Example

The value of a statistical life (VSL) is an artificial construct that gives a defined group of identical people's aggregate willingness to pay for a reduction in the risk of death that would result on the average in one less death for the group.

Numerical example of VSL: Assume that each person in a group of 100,000 is willing to pay $50 in higher annual energy costs to reduce his or her own annual risk of dying from 5 per 100,000 to 4 per 100,000. There will be one less death in this group if the policy is undertaken. But we cannot know which of the 99,996 survivors would have died without the policy. The group has an aggregate annual willingness to pay of $5 million ($50 x 100,000) and one death was avoided. Thus, the value of this *statistical* life is $5 million.

number of lives saved and multiplying by the VSL, it is numerically equal to the sum of the estimates of WTP values for small changes in risk to each exposed person. In the example in Box 2-2, the WTP value is not correctly interpreted as a $5 million value for the single life saved (which is not identified) but rather a $50 value to each of the 100,000 people who experience a reduction of 1 in 100,000 in mortality risk. It is not known whose premature death is avoided, so the concept of VSL is inherently ex ante in that it is based on values of changes in risk rather than values of identified lives.

The VSL is a computational tool to assist in scaling WTP estimates for valuing different magnitudes of risk change. Its use in this way is based on an implicit assumption that WTP to reduce risk is the same per unit of change in risk regardless of the cause or context of the risk.

There are many unanswered questions about how the VSL might vary in different contexts and in populations with different characteristics. One important and controversial question is whether WTP for mortality-risk reduction is expected to be proportional to remaining life expectancy. If it is, it may be more appropriate to use a monetary *value of a statistical life year* (VSLY) (see Box 2-3) or to adjust the VSL to reflect preferences of people at different ages rather than use the same VSL in all cost-benefit analyses. Evidence regarding that question and other factors concerning the estimation of the VSL are discussed in Chapter 5.

The choice between using a VSL or a VSLY to calculate mortality-risk reduction benefits is also related to whether mortality-risk reductions are expressed as reductions in lives saved or as reductions in life years saved. A calculation of lives saved is a measure of the number of people in a population who would have died in a given period but whose deaths are postponed because of

BOX 2-3 Definition of Value of a Statistical Life Year in Relation to Value of a Statistical Life, and Numerical Examples

The value of a statistical life year (VSLY) is another artificial construct, which gives a group of identical people's aggregate willingness to pay to extend the life of one person in the group by 1 y. In theory, the VSLY could be estimated directly; in practice, it is usually derived from an estimate of the value of a statistical life (VSL).

Numerical examples of VSLY: In the simplest case of a zero discount rate and a constant VSLY, the VSLY is the VSL divided by the number of life years saved by the reduction in risk (calculated from the remaining life expectancy of the people at risk). Thus, using the numerical example for the VSL from Box 2-2, if the average remaining life expectancy for the group with the VSL of $5,000,000 is 40 y, the VSLY equals $125,000 ($5,000,000 divided by 40). In practice, the values for future years are presumed to be discounted at some rate of time preference such that years further in the future are given less weight in current decision-making than years close to the present. For a discount rate of 5%, for example, a VSL of $5,000,000 implies a larger VSLY of about $291,000 if the VSLY is assumed to be the same for every remaining year. In this example, the VSL is interpreted as a present value of a stream of equal annual values over a 40-y period discounted at an annual rate of 5%. It is not required that the VSLY be a constant value for every year, although it is often assumed when a VSLY is calculated from a VSL.

the reduction in pollution exposure. A calculation of life years saved takes into account the remaining life expectancy of those whose deaths were prevented in a given period. For example, if the reduction in pollution exposure is such that 10 70-y-old people who would otherwise have died next year do not die then and if their average remaining life expectancy is 15 y, 10 lives and a total of 150 life-years are saved. Of course, the specific people whose deaths are prevented cannot be identified, and remaining life expectancy of any specific person is not known, but estimates can be made by using pollution-related risk estimates for age groups and using life tables that give average remaining life expectancies of people of different ages. This is discussed further in Chapter 4.

Proponents of using life years saved, rather than lives saved, as the unit of measure for mortality-risk reduction in cost-benefit analyses of environmental regulations have several arguments, including these:

- A measure of life years is intuitively appealing because lives are never "saved," rather, survival curves are shifted (that is, life expectancy is increased).
- Life years saved are commonly used in medical-care decision-making and health-care policy analysis.

• An intervention or program that provides a person a larger life expectancy increase is intuitively preferable to one that provides a smaller life expectancy increase.

The relative merits of the alternative methods for measuring and valuing mortality-risk changes are discussed in more detail in Chapters 4 and 5.

Role of Cost-Benefit Analysis[2]

As noted above, the use of economic analysis—of the costs and monetary benefits of an action—is not allowed in the setting of the NAAQS, but such analysis *is* conducted in three important arenas. The first is a set of retrospective and prospective analyses of the costs and benefits of the CAA; these analyses are required by Section 812 of the 1990 amendments to the act. Two have been conducted: the first looked backward for the period 1970-1990, and the second looked forward for the period 1990-2010 (EPA 1997b, 1999a, 2003a).

The other two uses of economic analysis are in response to a series of presidential directives, begun in the Ford administration and most recently codified in Executive Order 12866 during the Clinton era (with later interpretation by the George W, Bush administration's Office of Management and Budget, for example, in the recent Executive Order 13422). Executive Order 12866 mandates a regulatory-impact analysis (RIA) of the costs and benefits of any federal agency action expected to have a "significant economic impact" (that is, an economic impact expected to exceed $100 million). For the air-quality management process, that has two facets:

• First, an RIA is prepared for both a proposed and a final setting of each NAAQS. Although the law and recent court decisions (American Trucking Associations, Inc. v. U.S. EPA, D.C. Cir 97-1440 and 97-144) do not allow the results of this analysis to be used in setting the NAAQS, the RIA does provide public information about quantifiable costs and benefits of the proposal and inform public debate.

• Second, and perhaps more important, an RIA is prepared for every major EPA rule designed to reduce emissions of NAAQS pollutants or their precursors from mobile, stationary, and other sources. Although the RIA does not have a statutory role in EPA's setting of the standard, the CAA provisions that authorize actions to meet a standard normally do include cost among the criteria to be considered, so the cost and benefit analyses conducted in an RIA often are important in determining which actions to reduce emissions may be most effective in improving health at the lowest cost. Furthermore, Sunstein (2002) has argued

---

[2]The discussion in this section contains text that is excerpted or summarized from a document by L. Robinson (2007). The discussion is also informed by another document by L. Robinson (2004).

that a series of judicial decisions has created a default requirement that, "Unless Congress has clearly said otherwise, agencies will be permitted to take costs into account…and expected to balance costs against benefits in issuing regulations." (Sunstein 2002, p. 31)

The U.S. Office of Management and Budget (OMB) gives guidance for conducting RIAs—most recently in 2003 (OMB 2003). EPA has guidance for conducting the analyses that is more specific to the issues related to environmental regulation (EPA 2000a). OMB notes (OMB 2003, p. 2) that the motivation for the analyses is "to (1) learn if the benefits of an action are likely to justify the costs or (2) discover which of various possible alternatives would be most cost-effective."

OMB's guidance acknowledges that there are instances when important beneficial effects or costs of a regulation cannot be quantified or monetized. In such instances, OMB recommends a "threshold" or "break-even" analysis to evaluate the significance of nonquantified beneficial effects or costs. For instance, such an analysis would examine what the value of the nonquantified beneficial effects would have to be for the net benefits to be positive. For example, assume that a pollution-reduction program costs $10 million and has two benefits: benefits to the commercial forest industry that can be quantified and benefits to forest ecosystems that cannot be quantified. If the benefit to the commercial forest industry is estimated at $7 million, the benefit to forest ecosystems would have to be worth more than $3 million for the total benefits to exceed the costs.

EPA's guidance (EPA 2000a) provides more detail on what is to be included in quantified costs and benefits, how these are defined for cost-benefit analysis, and specific approaches that are used in the literature to quantify the types of costs and benefits that are typically affected by environmental regulations.

Both guidance documents include the following general recommendations:

- Conduct a comprehensive cost-benefit analysis for each regulatory alternative being assessed. This includes quantification and monetization of all costs and benefits, including those that might be considered ancillary (i.e., supplementary). When important costs or benefits cannot be quantified, a qualitative description of these costs or benefits should be included, as well as an assessment based on professional judgment of how important such costs and benefits might be.
- The methods used and assumptions made should be presented in a transparent way to the reader, so that the policy maker can understand the process by which the results were obtained.
- Uncertainty in the results should be communicated in a quantitative way, if possible, using techniques such as probability distributions, confidence intervals, or high, low bounds. The effects of key assumptions on the results should

also be presented, such as with sensitivity analyses that replace key assumptions with other plausible assumptions and demonstrate the effect on the results.

- In all cases, the cost-benefit analyses are intended to support decision making, not determine the decision. In some cases there are other statutory goals besides efficiency that are overriding. There are also other considerations such as the distribution of costs and benefits to various groups or sectors of society that should be taken into account.

Both of the guidance documents refer readers to standard texts on cost-benefit analysis and summarize the standard concepts that underlie this type of analysis. They note that the goal is to quantify both costs and benefits in terms of "opportunity cost." For benefits, this means a measure of what those who benefit would be willing to forgo to obtain the specified benefit. This is the definition of WTP, which is essentially a monetary measure of how much better off the benefiting population perceives themselves to be, assuming they have full information about what the benefits are. A fundamental point here is that the monetary measure of the value of the benefit is to be based on the value placed on the benefit by the group or sector receiving. It is not the value held by the policy maker or the experts. This is also why WTP estimates for changes in health are preferred to cost of illness estimates that reflect only medical costs and productivity losses.

EPA's Science Advisory Board, especially the Environmental Economics Advisory Committee, provides reviews and recommendations on drafts of EPA's guidance document for conducting RIAs, and provides specific advice on cost-benefit analysis. In addition, the Advisory Council on Clean Air Compliance Analysis was established under Section 812 of the 1990 Clean Air Act Amendments to provide advice on the cost-benefit analysis of the Clean Air Act that was mandated by Section 812. The Council has provided detailed advice on cost-benefit analysis for air pollution related assessments, and EPA has considered this advice in its RIAs as well as in the assessment of the costs and benefits of the Clean Air Act.

In *Circular A-4*, OMB (2003) discusses the quantification and economic valuation of mortality risk reductions. OMB does not require the use of any specific measure of effectiveness, such as lives saved or life-years saved, but encourages agencies to report results using multiple metrics that may provide different insights and perspectives.

> The OMB notes that [VSL and VSLY] are subject to continued research and debate and indicates that agencies should describe the limitations of their chosen approach. The *Circular* reports that the range of VSL estimates found in the literature is generally between $1 million and $10 million; as a result, regulatory agencies generally use values from within this range.

> In addition, *Circular A-4* discusses options for adjusting VSL estimates to reflect differences between the scenarios addressed in the research literature and the specific regulatory scenarios being assessed…It includes cautions on the application of age adjustments [for VSLs] and suggests the use of larger VSLY estimates for older individuals [if VSLY measures are used]….
>
> In some of its regulatory assessments [in 2003 and earlier], the EPA presented sensitivity analyses [using different VSL estimates for different age groups] based on research suggesting that older individuals are willing to pay less for life-saving interventions than younger adults (e.g., Jones-Lee 1989; Jones-Lee et al. 1993).
>
> In response, the OMB issued a memorandum advising agencies against adjusting the VSL for age (Graham 2003). This memorandum suggested that more recent research (ultimately published in Alberini et al. 2004a) did not fully support the VSL age adjustment found in earlier studies. It indicated that, when VSLY estimates are used instead of VSL, the yearly values are likely to be higher for senior citizens because "seniors face larger overall health risks from all causes and because they have accumulated savings and liquid assets to expend on protection of their health and safety" (Graham 2003, p. 2)...
>
> However, the guidance in this OMB memorandum, which was eventually incorporated into *Circular A-4*, does not necessarily eliminate the use of different values for younger versus older individuals. When VSLY estimates are applied, the total value of a risk reduction is equal to the product of the VSLY estimate and the discounted number of life-years saved. Unless the VSLY estimates for older individuals are large enough to compensate for the smaller number of life-years remaining, the use of VSLY estimates will result in lower values for older individuals (Robinson 2007, p. 286-287).

ENVIRONMENTAL PROTECTION AGENCY'S APPROACH TO ESTIMATING OZONE MORTALITY IMPACTS AND VALUING MORTALITY RISK REDUCTION[3]

Several examples show how EPA's benefits assessments include quantified and monetized ozone health benefits. All of these have included several morbidity endpoints in the primary estimates. Ozone mortality has been quantified as a sensitivity analysis based on available daily time series estimates of the

[3] The discussion in this section contains text that is excerpted or summarized from a document by L. Robinson (2007). The discussion is also informed by another document by L. Robinson (2004).

effect of ozone on mortality risk and using the VSL that has been used by EPA for PM mortality.

These have included the

- 1997 NAAQS RIA (EPA 1997a)
- Costs and Benefits of the CAA (EPA 1997b; prospective 1999a)
- Tier 2 Motor Vehicle Emission Standards RIA (1999b)
- Clean Air Interstate Rule (CAIR) RIA (EPA 2005a) and
- Proposed Small Engines Emission Rule (EPA 2007i).

EPA has prepared a RIA to accompany their proposal for revising the NAAQS for ozone (EPA 2007b). In that analysis, EPA has included calculations of reductions in mortality associated with reductions in ozone concentrations using alternative approaches and has presented all the results side-by-side. One calculation was based on the results of a multi-city study by Bell et al. (2004). This study provided the basis for the "primary" estimates of mortality reductions in EPA's earlier risk assessment. A second set of calculations was based on the three meta-analyses of ozone related mortality studies. These results gave mortality reduction estimates about 4 to 5 times larger than the estimates based on Bell et al. (2004), and that were relatively close to one another. The third calculation excluded any mortality associated with ozone and counted only the morbidity health outcomes. In addition to the health benefits of reductions in ozone, EPA calculated the health benefits of $PM_{2.5}$ reductions that would result from control strategies expected to be used to meet the ozone standards, which would reduce precursors that cause formation of both ozone and $PM_{2.5}$.

EPA's current primary approach for economic valuation of mortality risk reductions is a variation on their long-standing approach of using the same VSL for all annual mortality reductions. Starting with the RIA for the Clean Air Interstate Rule (EPA, 2005a), EPA has been using a central VSL that is a midpoint between results obtained in two meta-analyses of the wage-risk literature (described in Chapter 5). This is a VSL of $5.5 million in 1999 dollars and at 1990 income levels. When adjusted to expected income levels in 2006, this is about $6.3 (still in 1999 dollars). This VSL reflects the Agency's estimates of the individuals' willingness to pay (WTP) for small reductions in the risk of premature death for the population at large. EPA's current selection of VSL from the literature is similar to, but somewhat lower than the mean VSL EPA used earlier that was based on a review of a number of both revealed preference and stated preference studies (EPA 2000a).

In their primary benefits estimates, EPA applies the VSL to all lives saved regardless of the age or health status of the population experiencing the change in mortality risk and regardless of the cause of the mortality risk change. Although there is some expectation that willingness to pay for mortality risk reductions may vary with the characteristics of the population affected or with the context of the risk change, EPA has concluded that there is insufficient empirical information available upon which to base adjustments for these factors at this time. EPA's SAB (2007) has agreed with this conclusion regarding the state of the currently available literature. Following advice from the SAB, EPA adjusts its base VSL estimates for expected future real income growth and for any expected time lags between change in long-term pollution exposures and mortality risk reductions (often referred to as cessation lags). EPA included several different sensitivity analyses of the effects of age adjustments (adjusting VSL and/or applying VSLY estimates) in several of its reports prior to the development of OMB's *Circular A-4,* which cautions against making age adjustments to the VSL (Robinson 2007). These sensitivity analyses were included, for example, in the retrospective and prospective assessments of the costs and benefits of the Clean Air Act (EPA 1997b, 1999a) and in the RIAs for the heavy-duty diesel rule (EPA 2000b) and the Clear Skies legislation proposal (EPA 2003b, 2006b).

> "For example, for the heavy-duty diesel rule (EPA 2000c), the EPA used VSL age adjustments based on Jones-Lee (1989) and Jones-Lee et al. (1993) in sensitivity analysis, which reduced its primary benefits estimate by 10 or 40 percent, depending on the adjustment factor applied. In a sensitivity analysis for regulations addressing emissions from large spark ignition engines (EPA 2002), the agency used a more complicated approach that reflected initial results from the work of Alberini et al. (2004) as well as the adjustment factor from Jones-Lee (1989)" (Robinson 2007, pp. 290-291).

In this case, EPA derived different estimates of VSLY for younger and older age groups from selected VSL estimates. The result was a higher VSLY for the older age group. They applied these VSLY estimates of life-years saved for each age group. The net effect was a lower value for risk reduction for the older age group because the smaller remaining life expectancy more than offset the higher VSLY.

As part of the process of updating their Guidelines for economic analysis, EPA asked SAB to consider again the question of whether monetary values for mortality risk reduction in cost-benefit analyses should be adjusted for differences in the life expectancy of those at risk. The SAB (2007) concluded that economic theory provides indeterminate results about how remaining life expectancy may affect a person's WTP for mortality risk reduction and that this question must be addressed empirically. They further concluded that results in the empirical literature are also insufficient at this time to provide a basis for making quantitative adjustments to VSL according to the age or remaining life expectancy of the population at risk. In addition, they concluded that the empirical literature provides no support for the assumption implicit in the use of a constant VSLY that WTP for mortality risk reduction is proportional to remaining life expectancy. They recommended continued use of a constant VSL, with encouragement to EPA to fund more research to address the questions of how VSL varies with age and other population and risk characteristics.

OVERVIEW OF OTHER APPROACHES

Other Federal Agencies

Other federal agencies do not deal with ozone mortality in their regulatory analyses, but they often deal with other kinds of mortality risk.

> Other agencies promulgate fewer economically significant rules that require valuing the risk of premature mortality. Between October 2003 and September 2005, four agencies (in addition to the EPA) prepared final rules with quantified health and safety benefits that were reviewed by the OMB (OMB 2005, 2006). These agencies included the Food and Drug Administration (FDA) and the Centers for Medicare and Medicaid Services (CMS) in the Department of Health and Human Services (HHS), as well as the National Highway Traffic Safety Administration (NHTSA) and the Federal Motor Carrier Safety Administration (FMCSA) in the Department of Transportation (DOT). An earlier review, covering the period between January 2000 and June 2004, reported similar patterns in agency promulgation of major health and safety rules (Robinson 2004)" (Robinson 2007, p. 293).

Agencies of the Department of Health and Human Services (Food and Drug Administration and Centers for Medicare & Medicaid Services)

> The FDA does not provide formal internal guidance for economic analysis, but it applies a similar approach across many of its rules. For premature mortality, the agency often uses a VSL estimate of $5million, without specifying a dollar year, and occasionally provides alternative estimates using higher or lower values [see, for example, 68 Fed. Reg. 41434 [2003], 69 Fed. Reg. 9120 [2004]; 70 Fed. Reg. 33997 [2005]]. This estimate is roughly in the middle of the $1 million to $10 million range cited in *Circular A-4* (OMB 2003). The FDA rarely adjusts its VSL estimates for scenario differences, although it has addressed cessation lag (e.g., in its trans-fat rule, FDA 2003), and added the cost of cancer treatment ($25,000) and an adjustment for psychological factors ($5,000) to the VSL for a rule on X-rays [(70 Fed. Reg. 33997 [2005]]. Thus, while its base VSL estimates are similar to those used by the EPA, the values ultimately applied by the FDA may be quite different because of the income growth and other adjustments made by the EPA. A few FDA analyses have presented alternative estimates of the value of mortality risk reductions using VSLY as well as VSL estimates [e.g., 68 Fed. Reg. 41434 [2003]] (Robinson 2007, p. 293).

FDA sometimes uses VSLY estimates to quantify monetary values for changes in morbidity following the methods developed by Mauskopf and French (1991); these estimates are derived from VSL estimates in the literature assuming constant VSLY for remaining (discounted) life expectancy.

> Another HHS agency, CMS, develops few economically significant rules with health and safety impacts; most of its programs involve transfers (e.g., from taxpayers to Medicare and Medicaid recipients) and hence are not subject to the OMB requirements for regulatory analysis. In its immunization rule (70 Fed. Reg. 58835 [2005]), CMS applies the same VSL estimate as FDA ($5 million), noting that it is roughly the mid-point of the range of values suggested by OMB (Robinson 2007, p. 294).

Agencies of Department of Transportation (National Highway Traffic Safety Administration and Federal Motor Carrier Safety Administration)

Both the NHTSA and the FMCSA rely on the DOT guidance for their base VSL estimates. In contrast to EPA and the HHS agencies, the DOT agencies primarily address injury-related accidental deaths rather than deaths from illness.

> The DOT currently recommends the use of a $3.0 million VSL—noting that this value is imprecise and should be used as "a guide for thoughtful decision-making" (DOT 2002, p. 1). Its approach is based largely on the results of Miller (1990), with adjustments for inflation and newer studies. Miller's 1990 estimates vary from those used by the EPA because he applies different criteria to determine which studies to include, and adjusts the results to address certain limitations of the studies. The DOT indicates that it continues to review the literature and consider whether changes to this value are needed (DOT 2002) (Robinson 2007, p. 294).

California

The state of California conducted a benefits assessment during the process of making their latest revisions to the state ambient ozone standard in 2004. They selected central, low and high estimates of the percentage change in mortality association with daily fluctuations in ambient ozone concentrations, based on available daily time-series studies. The entire selected range was above zero, implying a 100 percent probability of a nonzero effect of ozone on mortality. Their results were not used to help set the standard, but to assess what the health benefits to the California public would be if the alternative ambient standards were met compared to current concentrations. In a subsequent publication, Ostro et al. (2006) included a monetary valuation for the ozone-related mortality cases, using EPA's selected estimate of the VSL.

Canada

The Canadian government has also used benefits assessments to determine the quantity and economic valuation of health effects associated with air pollution in Canada and for programs to reduce air pollution. Judek et al. (2005) estimated the effects of all air pollutants on mortality rates in Canada, including estimates of the effects of ozone on mortality. The concentration-response function for ozone was drawn from a daily time-series study of 12 Canadian cities (Burnett et al. 2004), using a multi-pollutant model the authors estimated for Health Canada (not reported in the published paper) (Personal communication with D. Stieb, August 8, 2007). The concentration response function for ozone mortality used was about 0.084% (standard error of 0.014%) change in mortality per ppb change in daily high-hour ozone. For monetary valuation of air pollution related mortality, Health Canada has been using a range of VSL estimates based on the WTP literature in Canada and the United States. These VSL estimates have been adjusted downward somewhat based on results from a relatively early stated preference study (Jones-Lee et al. 1985) that showed a lower WTP for mortality risk reduction for older subjects. VSL estimates from the working age population were thus adjusted downward to reflect the age distribution of non-accidental causes of mortality that are associated with air pollution. Health Canada is currently planning to update their procedures for selecting values for mortality risk reductions.

Europe

There have been several efforts in Europe to estimate both the mortality impacts of exposure to ozone, and to value those impacts. These have included the ongoing work of the European Union's ExternE program, as well as the July 2007 Clean Air Strategy of the United Kingdom (DEFRA 2007). While the general approach for estimating impacts and benefits is very similar to that followed in the United States (as described in Figure 1-1 in Chapter 1), there are three areas where these efforts have differed from normal EPA practice to date:

- First, these analyses have fairly unambiguously accepted from the epidemiologic literature that there are robust associations of mortality with ozone exposure and thus have used European time-series studies of these effects to estimate health impacts of current exposure, and the potential reduction in those impacts from regulatory actions.
- Second, these analyses have generally included an analysis of life-years lost, in the case of the UK including estimates of both life years and numbers of deaths (DEFRA 2007), while in the case of ExternE, rejecting the estimation of numbers of deaths and relying solely on life years lost (Bickel and Friedrich 2005).

• Third, the economic valuation estimates used in this ExternE analysis are from recent stated preference studies conducted in France, Italy, and United Kingdom (Alberini et al., 2006b). Their selected central estimate of EUR 50,000 per life year was derived from the median WTP value for the 5 in 1,000 mortality risk change over 10 years (median VSL was about EUR1.1 million). The ExternE authors converted this 10-year mortality risk change to its equivalent in increased life expectancy for each age/gender cohort and calculated the VSLY implicit in the WTP responses.

THE MAJOR QUESTIONS

As this overview has suggested, there is a broad and deep literature documenting a range of health effects from exposure to ozone, and this has had an important role in the setting and implementation of NAAQS for ozone for over 35 years. In considering whether and how the recent studies of ozone and mortality should be incorporated in estimating benefits going forward, the Committee focused on two primary areas of questions:

• *The Robustness of the Ozone Mortality Studies.* The estimation from epidemiologic evidence of health impacts, and the economic value of those impacts, requires confidence, as noted above, that ozone exposures can adequately be separated from other exposures in the epidemiology studies and that the epidemiology has adequately controlled for possible confounding factors. Although a number of others in Europe and California have made the judgment that these studies are adequate, the Committee was charged to make an independent determination on these questions. These questions are addressed in detail by the Committee in Chapters 3 and 4.

• *The Appropriate Methods for Estimating the Value of Potential Ozone Mortality Benefits.* If the epidemiology studies provide adequate evidence for the estimation of mortality impacts, there are important questions about the appropriate methods for estimating the value of such impacts that are described initially above, and are addressed in detail in Chapter 5.

3

Ambient Ozone and Related Pollutants

Ozone is a naturally occurring compound that is found in the troposphere (the layer of the atmosphere next to Earth's surface) and, at greater concentrations, in the stratosphere (the layer about 10-50 km above the surface). In the stratosphere, ozone absorbs and thus reduces the amount of potentially harmful ultraviolet (UV) radiation. Ozone in the troposphere also absorbs UV radiation and plays a role in the degradation of toxic compounds. More important, however, tropospheric ozone is responsible for various deleterious effects on humans and other organisms that are exacerbated by increased ambient ozone concentrations. Preindustrial tropospheric background ozone concentrations were about 10 ppb, but they have increased to about 30-40 ppb (e.g., Volz and Kley 1988; NRC 1991; Finnlayson-Pitts and Pitts 2000) because of emissions and atmospheric chemistry, as discussed below. In and around urban areas in the United States, ambient ozone concentrations rise much further, to about 160 ppb in Houston and 180 ppb in Los Angeles. Although such concentrations may seem extreme, they are substantially reduced from the concentrations of more than 400 ppb experienced in the 1970s (e.g., NRC 1991).

OZONE FORMATION AND TRANSPORT

We must understand the formation and transport of tropospheric ozone if we are to be able to interpret health-benefits assessments of ozone and conduct new ones, inasmuch as the interpretation of epidemiologic analyses must consider ozone's spatial and temporal patterns and their relationships with human exposure. Processes that affect ozone dynamics also affect other pollutants, such as components of particulate matter, and lead to what can be strong correlations between pollutant concentrations. This effect hinders our ability to determine the health effects of single pollutants that are present in a mixture. That is critical because a study may falsely implicate a pollutant that is not responsible for most of the health effects found, and strategies designed to reduce that pollutant may be ineffective in reducing the pollutants that *are* responsible for the health effects. An important question is whether that is the case in the approach to ozone and airborne particulate matter (PM), of which the latter seems to have greater health effects (see Chapter 2).

Most ozone in the troposphere is not directly emitted to the atmosphere (although there are minor sources of such ozone, including some indoor air cleaners) (e.g., NRC 1991). Rather, it is formed from a complex series of photochemical reactions of the primary precursors: nitrogen oxides (NO_x where $[NO_x]$ indicates the sum of [NO] and $[NO_2]$), volatile organic compounds (VOCs), and to a smaller extent other pollutants, such as carbon monoxide (CO) (Figure 3-1) (e.g., NRC 1991). The specific reactions that form most of the tropospheric ozone are the photolysis of NO_2 followed by the combination of the released oxygen atom with the abundant oxygen molecules (O_2):

$$NO_2 + \text{sunlight} \rightarrow NO + O \qquad [3\text{-}1]$$
$$O + O_2 \rightarrow O_3 \qquad [3\text{-}2]$$

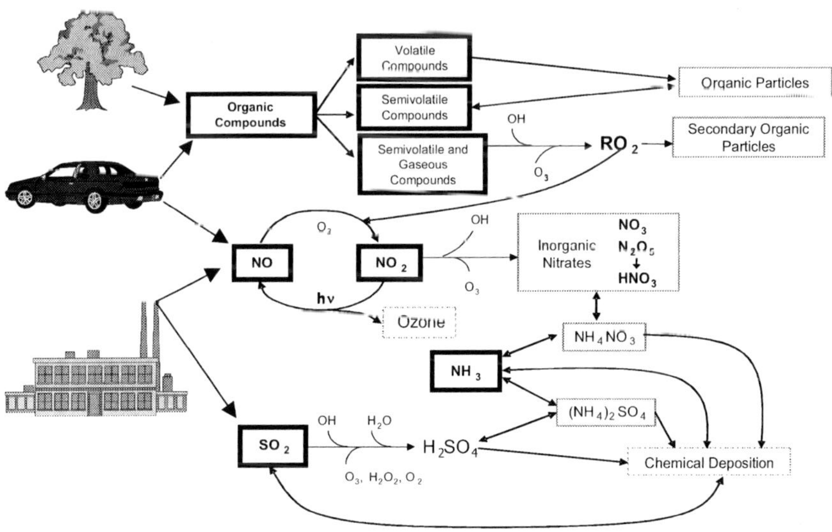

FIGURE 3-1 Source and chemical links between ozone and PM formation. Major precursors are shown in boxes with thick sides. Secondary particle components are shown in boxes with thin solid sides. Mobile sources (cars, trucks, and off-road vehicles) and plants are major sources of VOCs, and mobile sources and electricity-generating units are dominant sources of NO_x, but myriad smaller sources also contribute. Trace species, such as OH, are crucial to the formation of ozone, sulfate, nitrate, and organic-carbon particulate matter. Ozone also leads to the oxidation of SO_2 and NO_2. Biologic activity and fertilizer use dominate ammonia (NH_3) emissions. Source: Modified from NARSTO 2004. Reprinted with permission; copyright 2004, Cambridge University Press.

VOCs enter the picture by their reactions with the hydroxyl radical (OH) and other oxidants, which lead to the formation of NO_2 from NO, shown in a highly simplified form as

$$VOC + OH \rightarrow RO_2 \quad [3\text{-}3]$$
$$RO_2 + NO \rightarrow NO_2 + OPs \quad [3\text{-}4]$$

In Reactions 3-3 and 3-4, RO_2 is an oxygenated radical (there are many such species in the atmosphere), and OPs are oxygenated organic products, including aldehydes, ketones, acids, and condensable species that lead to secondary organic aerosol (SOA) formation. Although a small fraction of ground-level ozone is transported from the stratosphere, the reactions shown above (summarizing hundreds of reactions and species) are responsible for most ambient ozone, particularly on high-ozone days in urban areas (Fiore et al. 2003). Thus, ozone air-quality management strategies concentrate on reducing emissions of the ozone precursors. However, the nonlinearity of the relationship between ozone and its precursors (Figure 3-2) complicates decisions about which precursors should be controlled and to what degree.

Because reactions that form ozone are driven by sunlight, ambient ozone concentrations exhibit both diurnal variation (they are typically highest during the afternoon, as seen in Figure 3-3a) and marked seasonal variation (they are highest in summer, as seen in Figure 3-3b). Ambient concentrations are highest during hot, sunny summer episodes characterized by low ventilation (a result of low winds and low vertical mixing). Diurnal variability is enhanced, particularly in urban areas, by reactions between ozone and fresh emissions of NO (such as that from automobiles):

$$NO + O_3 \rightarrow NO_2 + O_2 \quad [3\text{-}5]$$

(An additional, more complex set of reactions leads to local reductions in ozone in response to increased NO_x emissions.) The impact of this Reaction 3-5 is particularly marked after the sun sets and during the day in areas with high NO_x (Figure 3-2); it forces ozone concentrations down, even below background concentrations. NO-associated decreases tend to occur relatively close to the NO sources. Indeed, increased NO_x emissions can lead to local ozone reductions (for example, near major roads and near the stack in power-plant plumes), whereas ozone formation downwind is enhanced (Figure 3-4) (see also Ryerson et al. 2001). The relationship between VOC emissions and ozone concentrations is somewhat simpler, with higher emissions typically leading to higher ambient ozone concentrations, particularly in urban areas.

The long effective atmospheric lifetime of ozone, measured in weeks (IPCC 2001), leads to substantial regional and intercontinental transport, and increased global NO_x emissions result in steadily increasing background ozone concentrations (e.g., Finnlayson-Pitts and Pitts 2000). That is in stark contrast

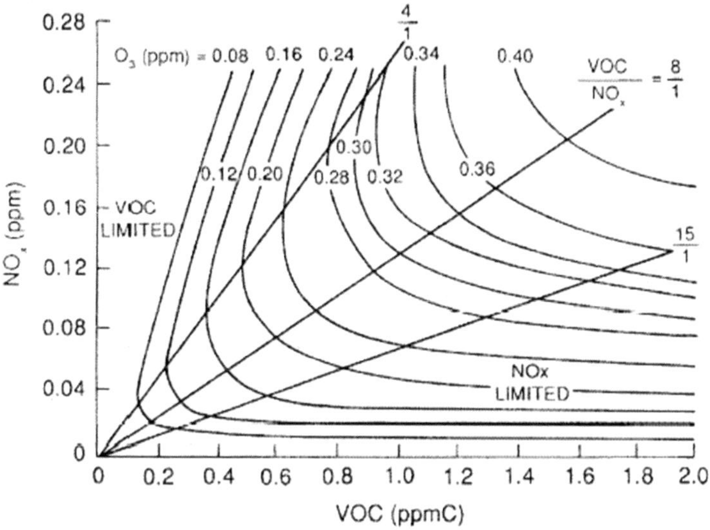

FIGURE 3-2 Ozone isopleth diagrams showing the nonlinear response of ozone to emissions of VOCs and NO_x and how they can vary from a city center to a downwind location. Source: Adapted from Dodge 1977.

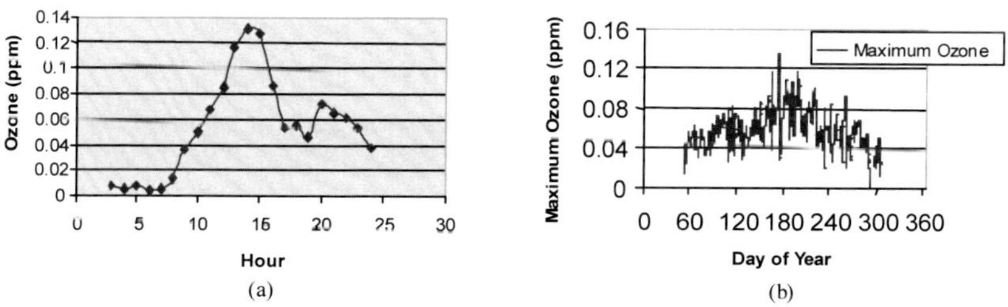

FIGURE 3-3 Ozone concentrations in Atlanta in 2006: (a) diurnal variation of ozone on July 22; (b) daily maximum.

with the initial view that ozone was an urban air pollutant that affected primarily the larger cities, such as Los Angeles (NRC 1991), and that it could be substantially mitigated by reducing VOCs. As pollution transport from high-NO_X-emitting areas, such as the Ohio River Valley, to downwind areas and cities became recognized as an important contributor to regional ozone, control efforts were redirected to reduce NO_X. Now industrialization of developing nations and recognition of the stability of ozone and its increased concentrations over the oceans have led to a global view of the problem. Global background concentrations are increasing, they are higher over the continents, and they are yet higher over and downwind of cities. As the ozone NAAQS is tightened, the regional concentrations approach the desired concentrations, providing very little room for the bump of ozone from individual cities.

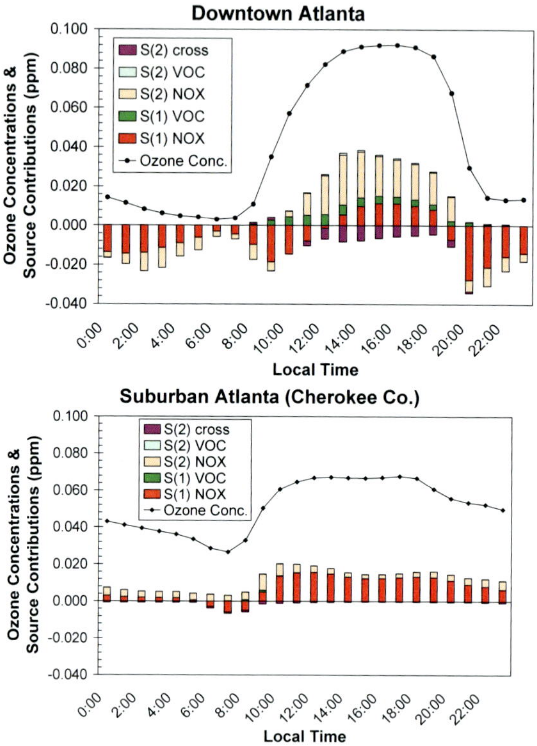

FIGURE 3-4 Simulated ozone concentration and sensitivity of ozone to Atlanta-area NO_x and VOC emissions for downtown Atlanta and a suburban location downwind. Bars in each graph illustrate how much local emissions change ozone concentrations in downtown and suburban Atlanta. Combined red and tan bars show how emissions of NO_x in the Atlanta area affect ozone. If bars are above zero (positive sensitivity), Atlanta NO_x emissions increase ozone in locations shown at that time. If bars are negative, such NO_x emissions lead to decreases in ozone at that location and time. Locally, increased NO_x emissions lead to decrease in ozone (negative sensitivity) through much of day downtown but generally increase ozone downwind in suburban area. VOC emissions increase ozone during daytime downtown (for example, green portion of bars) but have less effect at night and downwind. Gap between ozone concentration and sum of sensitivities shows magnitude of ozone transported from outside area. Reducing local NO_x emissions can decrease ozone peak but actually lead to an increase in 24-h average ozone on such a high-ozone day. On lower-ozone days, decreasing local NO_x emissions can increase ozone through more of day. Source: Cohan et al. 2005. Reprinted with permission; copyright 2005, American Chemical Society.

The amount of regional transport relative to local ozone production and the amount of NO_x emissions relative to VOC emissions lead to differences in the dynamics of ozone formation between cities (and in responses to control, as discussed below). Where VOC emissions are abundant (or NO_x emissions are low) and ozone concentrations are more homogeneous, ozone goes down relatively little at night, and weekend concentrations are similar to those in the rest of the week. That is particularly true in rural areas. In a place like Los Angeles, where ozone is largely produced locally and there are relatively high local NO_x emissions, ozone has a distinct diurnal trend, reaching a maximum in the afternoon and dropping toward zero in the evening and during the night. Ozone can increase on the weekend as NO_x emissions decline (for example, because of decreases in traffic and construction). That is the "weekend-weekday ozone" effect (e.g., Chow 2003 and references therein).

Despite the continuing rise in NO_x emissions worldwide and increases in vehicle miles driven and electricity production from fossil-fuel combustion in the United States, ambient ozone concentrations in cities in the United States and other developed countries have generally decreased. The reductions have occurred as the result of decreases in VOC emissions, which tend to have a local effect on reducing ozone, and NO_x controls, which lead to a more regional reduction. In many cities, particularly in the eastern United States, ozone originates primarily from distant VOC and NO_x sources; it is formed as prevailing winds transport pollutants into the cities. Local VOC and NO_x emissions cause more ozone to be added (e.g., NRC 1991, 2004a and references therein), and this leads to regionally high ozone concentrations with a noticeable increase in and downwind of cities (Figure 3-4; also see Box 3-1).

OZONE MEASUREMENT

In the United States, ozone concentrations are typically reported as a "mixing ratio," the ratio of the number of molecules of ozone to the number of all molecules in the same volume of air. For example, the typical unit used for regulatory purposes is parts per million, although scientific studies often use parts per billion. The prior NAAQS was an average of 0.08 ppm over 8 h, which, with rounding, was effectively 84 ppb, whereas the current NAAQS is 0.075 ppm. Other nations often use units of mass per volume, such as micrograms per cubic meter.[1]

Ground-level ambient ozone has for many decades been one of the most well-characterized pollutants in the United States, especially in more populated areas where multiple monitors are used. For example, the Los Angeles Basin has over 30 ozone monitors. Ozone is also measured in rural locations and in na-

---

[1]Ozone at 1 ppb is about 2 $\mu g\ m^{-3}$ (within about 2% at standard temperature and pressure).

BOX 3-1 Policy Relevant Background

EPA defines the policy relevant background (PRB) as the level that a pollutant concentration would be in the absence of anthropogenic emissions from the United States, Mexico, and Canada (see Chapter 2). However, because of intercontinental transport of ozone and its precursor emissions, it is not feasible to directly measure the amount of ambient ozone attributable only to PRB sources. Therefore, modeling is the only viable approach to estimating a PRB. Observations in very remote places can be used to assess how well a model simulates chemistry and transport in clean regions. In the 2006 Criteria Document, EPA used results from the GEOS-Chem global air quality model (Fiore et al. 2003), with U.S., Canadian and Mexican anthropogenic emissions removed, to provide temporally and spatially varying PRB ozone levels (EPA 2006a). As applied, GEOS-Chem used a spatial resolution of 2° latitude by 2.5° longitude (about 200 km x 275 km). Criticisms of this approach include the model's coarse spatial resolution and reliance on the assumption that Canadian and Mexican emissions would be potentially removed by U.S. policies (e.g., Brauer et al. 2007). Further, some analyses of observations at remote sites suggest that the calculated PRBs are low (Lefohn 2006; Brauer et al. 2007), noting that the model may underestimate the impact of some natural processes.

Although the approach used by EPA in calculating the PRB was reasonable given the information available at the time, there is a need to address the criticisms levied in order to provide a better foundation for a similar exercise in the future. The inclusion of Canadian and Mexican emissions removal in the process is a policy decision, and the influence of that choice is uncertain. GEOS-Chem's coarse resolution is of some concern, because the effect of using a model with a coarse resolution is uncertain, though given the lack of local emissions substantial concentration gradients are not expected. Although comparison of simulated ozone levels at a remote site to those observed did not show significant bias (Fiore et al. 2003, Hudman et al. 2004), demonstrating that the simulated distributions of observed ozone agree with observations at a variety of remote sites would strengthen the foundation of the modeling approach(es) chosen for future implementation. In regards to how to scale ozone levels between current levels and those meeting various proposed NAAQS, an assessment of how ozone will respond to the alternate path(s) of emissions removal would be of interest (e.g., compare removing 50% of the NOx, versus 50% of the VOC versus 50% of both). Such information can be used in the uncertainty assessments conducted as part of the NAAQS review process.

tional parks, for example, as part of the IMPROVE network (Sisler and Malm 2000). In many locations, however, ozone monitors are operated only during the ozone season, which varies from place to place. During the ozone season, monitors provide nearly continuous measurement, although levels are typically reported hourly. Thus, information on ozone, both spatial and temporal, is relatively complete in contrast with information on other pollutants. The wealth of information on ozone can be contrasted with that on airborne fine PM with an aerodynamic diameter less than 2.5 μm ($PM_{2.5}$), which has been widely monitored for less than a decade and of which many of the measurements are for 24-h periods as opposed to hourly in the case of ozone.

Chemiluminescent monitors or UV absorption monitors are typically used for routine ozone measurements. A recent Environmental Protection Agency (EPA) analysis found that when averaged over 8 h monitor precision is better than 3%, which is why the ozone design value (for an emission-control program) determined from an area's monitoring data is within about 1.3 ppb (about 3 $\mu g\ m^{-3}$) (EPA 2007a). For regulatory purposes, data are rounded to the last figure in the level of the standard; thus, the previous standard of 0.08 ppm has led to an effective standard of 0.084 ppm (EPA 2007a). In other words, a measured value of 0.084 ppm would be rounded down to meet the 0.08 standard. This improved measurement accuracy has prompted EPA and its Clean Air Science Advisory Committee to recommend making the revised ozone NAAQS more precise, that is, adding an extra digit to the level of the standard. In March 2008, EPA set the level of the standard at 0.075 ppm (EPA 2008a).

SOURCES OF OZONE PRECURSORS AND OZONE CONTROL

There are myriad major outdoor sources of VOCs, including vegetation, solvent use, and mobile sources (Figure 3-5). Ambient sources of NO_x include fuel combustion (for example, in cars, trucks, construction equipment, factories, and power plants) and to a lesser extent biogenic activity (Figure 3-5). The distribution of sources and the other factors that contribute to ozone cause the highest ozone concentrations to be in or just downwind of larger cities, particularly cities that are sunny during summer, such as Los Angeles, Houston, cities in the Northeast Corridor, and Atlanta. However, smaller cities can experience ozone concentrations above the previous NAAQS of 0.08 ppm; with the promulgation of a tighter ozone standard, even more areas are expected to be out of attainment of the NAAQS for ozone (see Figures 2-1 and 2-2 in Chapter 2).

Early ozone-control programs concentrated on VOC-emission reductions because they were thought to be cost-effective, decreased concentrations of many additional air toxicants, and, unlike control of NO_x, were unlikely to raise ozone concentrations. However, as the regional nature of the ozone problem became apparent, the focus shifted to NO_x-emission controls, which had a greater influence on regional ozone concentrations. For example, the recent Clean Air Interstate Rule (CAIR) (EPA 2007h), which is designed to reduce ozone and $PM_{2.5}$, focuses on reducing NO_x and SO_2 from electricity-generating units and other large sources in 28 eastern states and the District of Columbia.

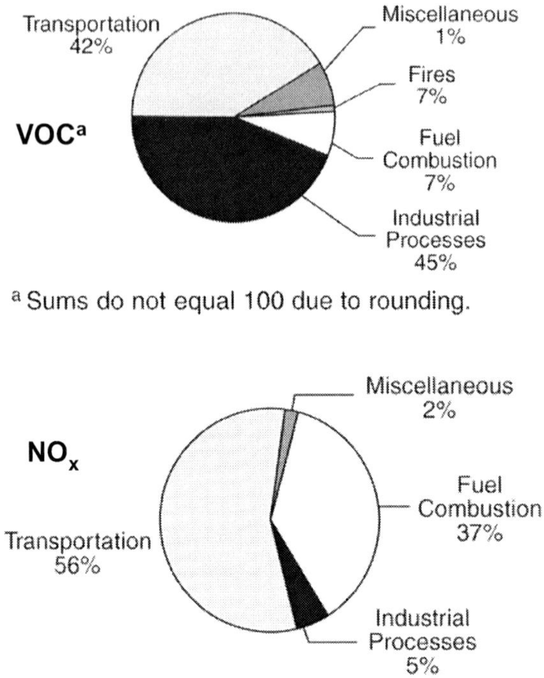

FIGURE 3-5 Anthropogenic sources of 2002 ozone-precursor emissions of VOCs and NO_x. Biogenic sources are not included but dominate nationwide VOC emissions. Source: EPA 2003c, Figures 2-19 and 2-36.

Automobile standards, which originally limited VOC and CO emissions, now include limits on NO_x and PM emissions. If the imposition of CAIR, tighter automobile and truck emission standards, and other controls are effective to the extent expected by EPA, they will lower ozone substantially in the near future (Figure 3-6). Those controls would reduce NO_x by about 42% and anthropogenic VOCs by 28% by 2020 (Woo et al. 2006). Given that those steps are not likely to bring all areas into attainment, state improvement plans will need to include further controls.

RELATED POLLUTANTS

In assessments of the mortality risks associated with ozone exposure, it is important to consider how emission sources, meteorology, and the chemistry of ozone formation affect the formation of other pollutants of concern, including components of particulate matter (for example, the sulfate fraction of $PM_{2.5}$) and other gases, such as aldehydes, and acids (Figure 3-1). The sharing of many of

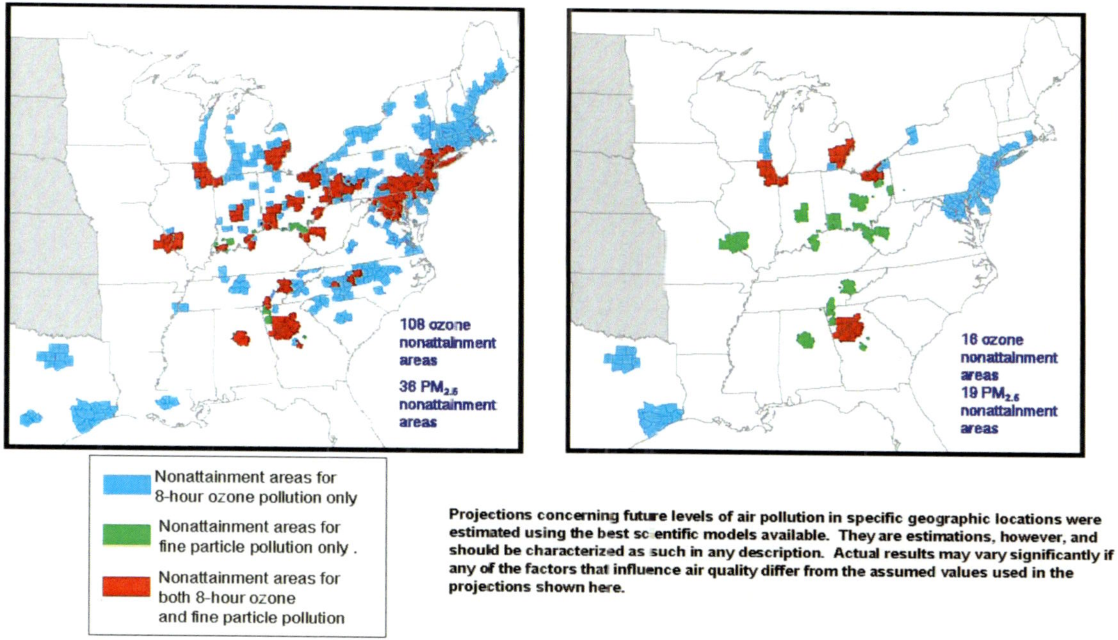

FIGURE 3-6 *Left*, ozone and $PM_{2.5}$ nonattainment areas in the eastern United States in 2006. *Right*, implementation of CAIR and other controls is expected to bring future ozone concentrations in many areas in the eastern United States into attainment of the previous 0.08-ppm ozone NAAQS. Source: EPA 2005c.

the sources and the intertwining of chemical and meteorologic processing lead to strong correlations between ambient ozone concentrations and concentrations of the other pollutants, such as sulfate and particulate organic carbon in summer (see Tables 3-1a, b, c and Table 3-2). The strong linkages and the fact that the other pollutants may have health effects are captured in a regulatory context in

TABLE 3-1a Day-to-Day Correlation Between Air Pollutants in Boston, MA, by Season (Summer Nonshaded, Winter Shaded), 1999-2004[a]

	O_3	$PM_{2.5}$	SO_4^{2-}	BC	PN	NO_2	CO	SO_2
O_3	1.00	0.55[b]	0.59[b]	-0.06	-0.34[b]	0.10	-0.03	0.01
$PM_{2.5}$	-0.62[b]	1.00	0.74[b]	0.43[b]	-0.37[b]	0.27[b]	0.36[b]	0.11
SO_4^{2-}	-0.43[b]	0.78[b]	1.00	0.34[b]	-0.31[b]	0.25[b]	0.14	0.13
BC	-0.60[b]	0.67[b]	0.64[b]	1.00	0.11	0.55[b]	0.39[b]	0.17
PN	-0.13	-0.04	0.10	0.10	1.00	0.25[b]	0.03	0.22[b]
NO_2	-0.56[b]	0.60[b]	0.57[b]	0.64[b]	0.33[b]	1.00	0.54[b]	0.32[b]
CO	-0.58[b]	0.50[b]	0.48[b]	0.72[b]	0.23[b]	0.59[b]	1.00	0.33[b]
SO_2	-0.42[b]	0.64[b]	0.50[b]	0.47[b]	0.32[b]	0.63[b]	0.60[b]	1.00

[a]Pearson correlation coefficients. Summer defined as May-August, winter as November-February. BC = black carbon. PN = particle number. [b]indicates p-value < 0.0001.

TABLE 3-1b Day-to-Day Correlation Between Air Pollutants in St. Louis, MO, by Season (Nonwinter Clear, Winter Shaded), April 2001-September 2002[a]

	O_3	$PM_{2.5}$	SO_4^{2-}	NO_{3-}	BC	OC	NO_2	CO	SO_2
O_3	1.00	0.39[b]	0.37[b]	-0.54[b]	0.02	0.30[b]	-0.10[c]	-0.21[b]	0.06
$PM_{2.5}$	-0.29[c]	1.00	0.72[b]	0.06	0.50[b]	0.60[b]	0.23[b]	0.24[b]	-0.03
SO_4^{2-}	--	--	1.00	--	0.47[b]	0.69[b]	0.18[c]	0.22[c]	0.02
NO_{3-}	--	--	--	1.00	0.07	-0.06	0.28[b]	0.32[b]	-0.01
BC	-0.39[b]	0.61[b]	--	--	1.00	0.77[b]	0.69[b]	0.52[b]	0.03
OC	-0.20	0.68[b]	--	--	0.92[b]	1.00	0.47[b]	0.42[b]	0.17[c]
NO_2	-0.23[c]	0.53[b]	--	--	0.70[b]	0.70[b]	1.00	0.49[b]	0.16[c]
CO	-0.35[b]	0.53[b]	--	--	0.86[b]	0.79[b]	0.53[b]	1.00	0.05
SO_2	-0.02	0.30[b]	--	--	0.29[c]	0.27[c]	0.35[b]	0.25[c]	1.00

[a]Pearson correlation coefficients. Winter defined as November-March, nonwinter as April-October. Data from St. Louis Supersite. $PM_{2.5}$ measured with beta-attenuation gauge monitor. BC = black carbon. OC = organic carbon. Sulfate and nitrate data not available during winter 2001-2002. Sample sizes for pairwise comparisons generally above 70 in winter and from about 110 to 390 in nonwinter (except 53 for SO_4^{2-} and OC comparisons in nonwinter). [b]indicates p-value < 0.0001. [c]indicates p-value < 0.05.

TABLE 3-1c Day-to-Day Correlation Between Air Pollutants in Los Angeles, CA, by Season (Nonwinter Clear, Winter Shaded), June 2002-December 2003[a]

	O_3	$PM_{2.5}$	SO_4^{2-}	NO_{3-}	EC	OC	NO_2	SO_2
O_3	1.00	0.08	0.32[c]	-0.08	-0.56[b]	-0.32	-0.38[b]	-0.07
$PM_{2.5}$	-0.32[b]	1.00	0.87[b]	0.78[b]	0.33[c]	0.49[b]	0.46[b]	-0.04
SO_4^{2-}	-0.08	0.79[b]	1.00	0.68[b]	-0.01	0.17	-0.05	0.09
NO_{3-}	-0.27	0.97[b]	0.82[b]	1.00	0.33[c]	0.44[c]	0.31[c]	-0.02
EC	-0.80[b]	0.65[c]	0.21	0.52[c]	1.00	0.87[b]	0.86[b]	0.28[c]
OC	-0.60[c]	0.82[b]	0.40[c]	0.69[b]	0.84[b]	1.00	0.89[b]	0.31[c]
NO_2	-0.64[b]	0.65[b]	0.41[c]	0.64[c]	0.87[b]	0.90[b]	1.00	0.21[b]
SO_2	-0.30[b]	0.28[c]	0.25	0.27	0.51[c]	0.41[c]	0.55[b]	1.00

[a]Pearson correlation coefficients. Winter defined as November-February, nonwinter as March-October. Data from EPA Speciation Network. EC, elemental carbon. OC, organic carbon. In winter, sample sizes for pairwise comparisons with particles 27 or 28 and for comparisons between gases, about 180; in summer, samples sizes for comparisons with particles above 60 and for comparisons between gases, 370. [b]indicates p-value < 0.0001. [c]indicates p-value < 0.05.

that the NAAQS is for ozone *and other* photochemical oxidants; ozone is used as an indicator of the presence of the wide array of photochemical oxidants present in ambient air. (Other photochemical oxidants include a wide variety of both organic and inorganic species, such as nitrogen dioxide, organic and inorganic acids [sulfuric, nitric, formic, etc.], organic and inorganic peroxides, other reactive oxygen species such as quinones, and many more. [see also Figure 3-1 for chemical linkages].) Similarly, epidemiologic studies of ozone-health associations tacitly include the potential exposure to co-occurring pollutants unless they control for such additional exposure. However, the complete suite of other pollutants has never been monitored sufficiently to be included in such studies. Many of the photochemical oxidants that would be expected to covary with ozone are hardly, if ever, monitored.

Correlations between ozone and other pollutants, both measured and unmeasured, have led to an awareness of the possible confounding of ozone-health effects analyses. As shown on Tables 3-1a-c, correlations between ambient ozone and other pollutants, such as PM, vary substantially by geographic location, season, and pollutant. For example, associations between ambient ozone and $PM_{2.5}$ differ substantially in the Western and Eastern United States, between summer and winter, and for different PM components. Also, correlations among ambient ozone and other pollutants have been shown to vary by averaging period. These results suggest that the potential for confounding of ozone-health effects also varies by these factors, as discussed in Chapter 4. Strong correlations between ambient ozone and its co-pollutants further suggest that it may be difficult to separate the effects of ambient ozone from its co-pollutants in epidemiologic studies based on using ambient concentrations as an estimate of exposure.

Of ozone's copollutants, ambient $PM_{2.5}$ has been of particular concern with regard to confounding because of its observed effects on health and the often strong, seasonally varying correlations between ambient concentrations of ozone and $PM_{2.5}$ (Tables 3-1 and 3-2). $PM_{2.5}$ is a mixture of many compounds that originate from a variety of primary emission sources and reactions in the atmosphere. Not only do the formation and fate of specific compounds vary temporally and spatially but health assessments suggest that the components have different health effects (e.g., Laden et al. 2000; Peel et al. 2005; Thurston et al. 2005; Sarnat et al. 2008).

Strong summer correlations between ozone and $PM_{2.5}$ may be attributed to similar formation processes. For example, in New York and other eastern U.S. states (Figure 3-6), $PM_{2.5}$ in summer consists largely of sulfate, which originates primarily from the oxidation of emitted SO_2 by the hydroxyl radical (Figure 3-1, Reaction 3-6), the same molecule that leads to oxidation of VOCs:

$$SO_2 + OH \rightarrow \ldots \rightarrow H_2SO_4 \rightarrow \text{sulfate aerosol.} \qquad [3\text{-}6]$$

In winter, particulate nitrate formation from emitted NO_x can be important. It, too, is oxidized by the hydroxyl radical:

$$NO_2 + OH \rightarrow HNO_3 \rightarrow \ldots \rightarrow \text{nitrate aerosol.} \qquad [3\text{-}7]$$

In addition, some organic products (OPs) formed in Reaction 3-4 can lead to SOA production and thus also contribute to $PM_{2.5}$ concentrations. In contrast, ozone concentrations can be depressed in areas with high NO and NO_2 (such as areas close to highways). That also means that ozone concentrations can be lower in areas that have pollutants coming from substantial NO_x sources. For example, high concentrations of ultrafine particles (less than 100 nm in diameter) are found near freeways, which also have higher concentrations of CO and NO_x (Zhu et al. 2002; Sardar et al. 2004). However, ozone and $PM_{2.5}$ can correlate positively away from the source, for example, because of similar formation routes and precursor emission sources of ozone and sulfate (Figure 3-1).

The relation between ambient ozone and $PM_{2.5}$ can differ substantially between winter and summer, especially in areas with low winter sulfate, such as the eastern United States (Tables 3-1a,b,c). As a result of lower sulfate formation, primary $PM_{2.5}$ sources—such as motor vehicles, wind-blown dust, and combustion—contribute a larger fraction of ambient $PM_{2.5}$. Furthermore, secondary nitrates increase in winter because of a decrease in competition for ammonia (NH_3) by sulfate and because of the decrease in temperature, both of which

TABLE 3-2 Pollutant Correlations at Jefferson Street SEARCH Site, Atlanta, GA, 1999-2006[a]

	24-h O_3	1-h max O_3	8-h max O_3	24-h O_3 (O_3 season)	1-h max O_3 (O_3 season)	8-h max O_3 (O_3 season)
$PM_{2.5}$	0.31	0.52	0.47	0.60	0.74	0.73
SO_4	0.56	0.65	0.63	0.55	0.61	0.61
Organic carbon PM (OC)	0.02	0.24	0.19	0.55	0.70	0.70
24-h O_3	1.00	0.37	0.91	1.00	0.80	0.87
1-h max O_3	0.87	1.00	0.98	0.80	1.00	0.97
8-h max O_3	0.91	0.98	1.00	0.87	0.97	1.00
24-h O_3 (O_3 season)	1.00	0.80	0.87	1.00	0.80	0.87
24-h NO	-0.31	-0.06	-0.11	-0.07	0.25	0.23
24 h average NO_2	-0.13	0.12	0.07	0.16	0.45	0.42
24-h average PM_{10}	0.31	0.53	0.49	0.55	0.74	0.73

[a]This SEARCH site (Hansen et al. 2006) is the only location that has such a long record of daily speciated $PM_{2.5}$, PM_{10}, NO_x, and O_3 for such an analysis.

increase the stability of ammonium nitrate. At the same time, the decrease in solar radiation leads to lower hydroxyl-radical concentrations and reduced VOC oxidation and thus to lower NO conversion to NO_2. The primary NO emissions then scavenge ozone. Thus, on days when primary PM and NO are poorly dispersed, concentrations of both increase, and the increased NO lowers ozone (often to well below background concentrations). Ambient concentrations of ozone and $PM_{2.5}$ may not correlate or may even correlate negatively in winter (Tables 3-1a,b,c). The seasonal variability in the ambient ozone-$PM_{2.5}$ association suggests that the potential for confounding of ozone health effects by $PM_{2.5}$ also differs by season, with confounding most likely when they correlate strongly in summer (see also Bell et al. 2007).

The potential for confounding of ambient-ozone health effects by ambient $PM_{2.5}$ also probably depends on particle composition, which varies with location and season (Figure 3-7). As shown above, ambient ozone concentrations tend to correlate strongly and positively with ambient sulfate but not with ambient traffic-related elemental carbon, where a negative correlation is often found. Since studies have suggested that $PM_{2.5}$ health impacts may differ by component, these results indicate that confounding by specific $PM_{2.5}$ components, including sulfate, elemental carbon, metals, and secondary organics, should also be examined. It will be difficult to address such confounding with currently available data, however, because data on the $PM_{2.5}$ component have only recently been collected routinely in many sites. EPA's Speciation Trends Network (STN) for monitoring $PM_{2.5}$ components started in 2000, and individual monitors came on line at different times. Most monitors measure a standard set of characteristics, including $PM_{2.5}$ mass and various components (Peterson et al. 2000; Flanagan et al 2006). In the near future, new speciated measurements may provide the needed data, but, unlike ozone data, such data are now generally available only once every 3 or 6 d. To understand how short-term variation in $PM_{2.5}$ components might confound ozone-mortality associations, more frequent measurements may be needed. One other approach to avoiding the confounding by $PM_{2.5}$ is to consider the spatial distribution of the two pollutants, although they have similar spatial distributions (Figure 3-4b), which are tied to their similarities in sources, transport, and chemistry. Confounding of epidemiologic studies of ozone-mediated mortality and other health effects by $PM_{2.5}$, especially by such specific components of $PM_{2.5}$ as sulfate, presents formidable challenges to the interpretation of both time-series and spatial distributions. A further challenge to epidemiologic studies is that there are hundreds of gaseous and particle-bound ozone-related pollutants, including peroxides, other reactive oxygen species, and oxidized organics—most of which are not routinely monitored. The correlations between ozone and PM, as well as the lack of characterization of the variety of ambient pollutants that may also contribute to adverse health outcomes, presents a challenge to epidemiologic studies of PM effects and to studies of ozone effects because of the potential for confounding by many of the pollutants in the atmospheric mixture.

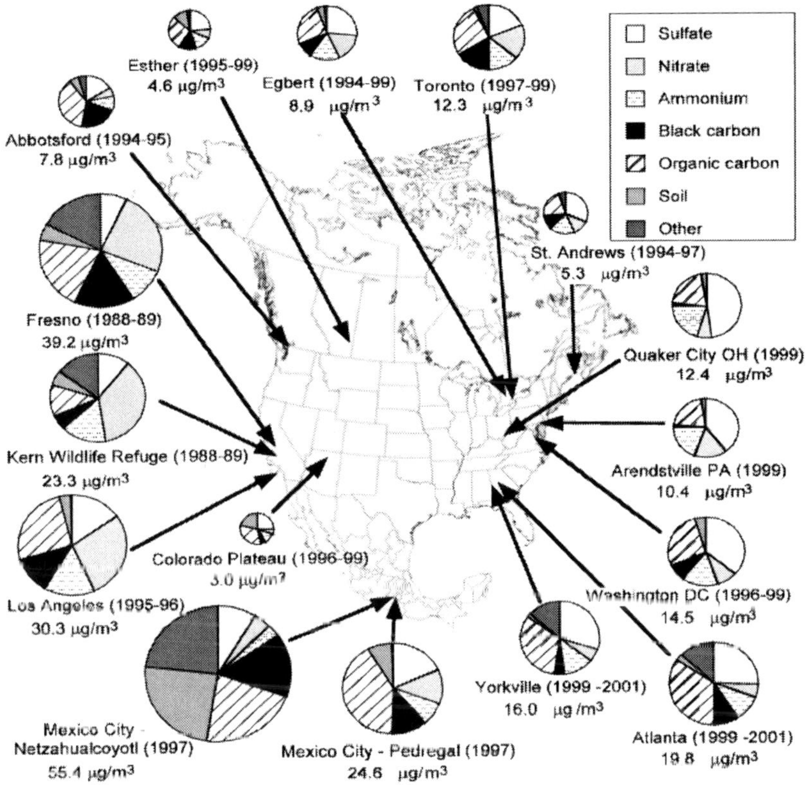

FIGURE 3-7 Composition of $PM_{2.5}$ in representative urban and rural locations. Urban sites are Toronto, ON, Canada; Washington, DC; Atlanta, GA; Mexico City, Mexico; Los Angeles, CA; and Fresno, CA. Averaging periods and average $PM_{2.5}$ mass are indicated. All sites have at least 1 y of sampling except Mexico City, for which average was determined for 14 d in 1 mo. More recent short-term measurements from December 1995 and January 1996 at Fresno and Kern Wildlife Refuge in California show lower $PM_{2.5}$ mass concentrations than displayed here but similar composition. Colorado Plateau data are averages of IMPROVE sites in Bryce Canyon, Canyonlands, Grand Canyon, Petrified Forest, Mesa Verde, and Zion National Parks. Source: NARSTO 2003. Reprinted with permission; copyright 2005, Cambridge University Press.

OZONE CONTROL

EPA uses morbidity and mortality risk estimates to assess the benefits of its pollutant-control programs and thus the benefits of reductions in specific emissions from specific sources. That approach has a subtle but profound effect on the metric chosen to link ambient ozone concentrations with mortality risk. Various studies (e.g., Bell et al. 2005; Levy et al. 2005) have considered different metrics to characterize short-term ozone exposure, including the 24-h average, the 8-h maximum (similar to the NAAQS), and the 1-h maximum (similar to the prior NAAQS). In meta-analyses of studies of the relationships between ozone and mortality (e.g., Bell et al. 2005), ozone concentrations were linearly scaled to the same metric to allow comparisons across studies; for example, a 24-h ozone average was used by Bell et al. (2005). Scaling of the 1-h maximum, the 24-h average, and the 8-h maximum ambient ozone concentrations is reasonable given their typically strong correlations for a given location, but it is not generalizable to ozone-control strategies in that the 24-h average and 1-h and 8-h maximum ambient ozone concentrations will respond differently to specific control strategies (Table 3-3) (Liao et al. 2008). For example, although measures to reduce ground-level NO_x emissions in a city, such as that from cars, may result in lower daily maximum ambient ozone concentrations on sunny stagnant days (for example, see Figure 3-8) (Cohan et al. 2005) in or downwind of such cities as Atlanta, Houston, and New York (Sillman et al. 1990; NRC 1991), the measures can also result in higher 24-h averages because of the higher nighttime ozone concentrations due to reduced ozone scavenging at night (as discussed above). Furthermore, on days that are cloudy or otherwise less conducive to ozone formation, NO_x-reduction measures can result in both higher peak and higher average ozone; that is, the sensitivities of ozone to NO_x controls are negative (Figure 3-8). Thus, strategies aimed at meeting the NAAQS (based on reducing the fourth-highest 8-h maximum) can lead to increases in average ozone level concentrations (Table 3-3).

This issue is directly related to assessments of the characteristics of ozone exposure that lead to death and other end points of concern. As shown in Table 3-2 and noted in various studies (e.g., Bell et al. 2004), various ozone metrics for data collected in a specific location tend to be highly correlated, so it may seem unimportant to distinguish between them; given the degree of correlation, it may be difficult for an epidemiologic study to do so. However, the choice of metric can be influential in determining benefits of control programs. For example, a program that lowers NO_x emissions could reduce peak ozone concentrations but raise average concentrations. A cost-benefit analysis based on mortality and average ozone could appear to show a negative benefit although the association with peak summertime concentrations appears to be positive. As discussed later, it is unknown which is more accurate, so evaluating the benefits of urban NO_x reduction is uncertain in both magnitude and sign. Complicating the process is that the association might depend on the metric of exposure, that is, whether the response is taken to have a threshold or not. In the recent NAAQS review process, EPA assumed a linear nonthreshold response (EPA 2007a).

Control programs for ozone will affect other pollutant concentrations because ozone precursors are also responsible for formation of other species; pollutants have many sources in common. The effects of these ozone-control programs on other species will vary in space and time and, as discussed above for ozone, can even change direction, depending on conditions. In general, reduc-

TABLE 3-3 Simulated Sensitivity of Annual Average Ozone to NO_x and Sensitivity of Fourth-Highest 8-h Average Ozone to NO_x[a]

	Atlanta	Chicago	Houston	Los Angeles	New York
2001 annual average ozone	-0.06	-0.11	-0.05	-0.07	-0.11
2001 fourth-highest 8-h average ozone	0.42	0.26	0.28	0.05	-0.29

[a]Negative sensitivity means that ozone will go up if NO_x emissions are reduced. Sensitivities provided are local (for example, in given conditions, how ozone will respond to emission changes) and are given as parts per billion per 1% change in domainwide NO_x emissions. As shown, annual average ozone goes up in each location, although fourth-highest 8-h averages typically are reduced. Source: Adapted from Liao et al. 2008. Reprinted with permission; copyright 2008, American Chemical Society.

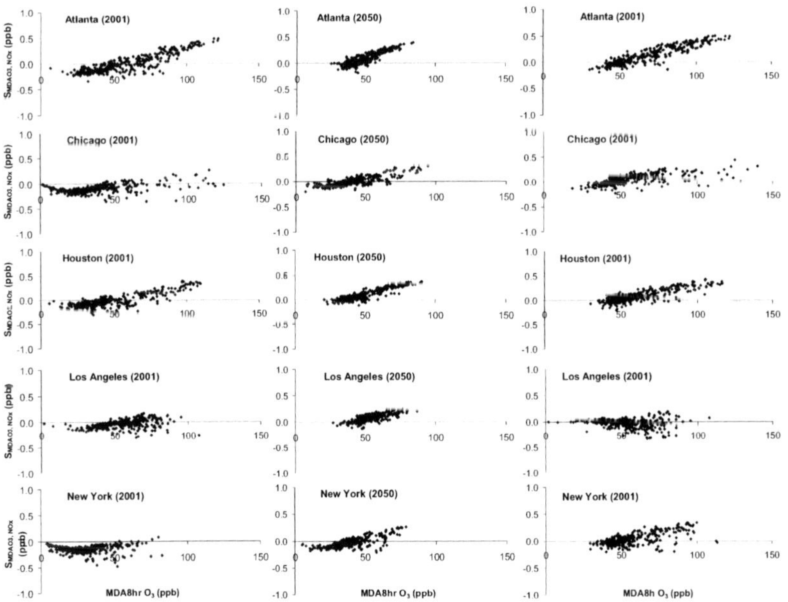

FIGURE 3-8 Calculated daily sensitivities of maximum 8-h averaged ozone in Atlanta, Chicago, Houston, Los Angeles, and New York regionwide changes in NO_x emissions and corresponding simulated concentrations. Shown are responses both at the city center and at the location of the regional maximum for 2001, and for the city-center case in 2050 in response to a 51% reduction in NO_x emissions. Results shown are for change in ozone in parts per billion per 1% change in domainwide NO_x emissions. Although highest 8-h ozone average may decrease with NO_x controls, lower peak daily concentrations can increase. Ozone formations at the location of the regional maximums tend to be more NO_x-limited. Source: Liao et al. 2008. Reprinted with permission; copyright 2008, American Chemical Society.

tions in NO_x will directly reduce nitrate formation (e.g., Russell and Cass 1986; Boylan et al. 2006), and there is evidence that they will slightly increase sulfate. For example, Liao et al. (2007b) found that the effect of a 1% reduction in NO_x emissions would be a 0.2-ppb reduction in average fourth-highest 8-h maximum ozone, a 0.01-$\mu g/m^3$ reduction in average annual nitrate aerosol, and a slight increase in average annual sulfate, with some areas experiencing small increases and others seeing decreases. Reductions in anthropogenic VOC emissions would reduce SOA formation, although the relationship between anthropogenic and biogenic VOC emissions and SOA formation is uncertain, and it would depend on which VOC species were being controlled (e.g., Robinson et al. 2007). Some organic compounds (such as benzene) are toxicants or (as in the case of formaldehyde) result in formation of toxicants in the atmosphere. Effects on toxic compounds will depend on the type of control and the source. For example, reducing automobile VOC emissions will probably reduce benzene, and changing paint solvents will affect the concentrations of other compounds. VOC controls will lead to minor changes in sulfate and nitrate concentrations (Napelenok et al. 2007).

Much less is known about how ozone-related controls will affect concentrations of reactive oxygen species (ROS), in large part because of the general lack of specific knowledge about ROS. The chemistry of oxidant species and the presence of their precursors suggest that reductions in VOC emissions would lead to reductions in organic ROS. If the ROS is associated with metals, the effect of ozone-related controls is less obvious, because the dominant precursor sources are less common and the chemistry not as well explored.

Some PM-related controls (such as NO_x controls, as discussed above) will affect ozone concentrations and (as in the case of control of fly ash) could reduce metal-associated ROS. SO_2 reductions will have a minor effect on ozone concentrations (Napelenok et al. 2007), and the direction of the response is uncertain (Dickerson et al. 1997; Jacobson 1998).

OZONE DYNAMICS AND MONITORING: IMPLICATIONS FOR HEALTH STUDIES

Time-series studies of ozone typically assess health effects in terms of ambient concentrations measured over time at a central monitoring site or at several sites in the study area. The use of ambient concentrations to estimate exposures is a source of considerable uncertainty in understanding ozone-mediated mortality with respect to both how well they indicate ozone exposures over time and space and how well ozone effects can be separated from effects of other pollutants or weather conditions. The relation of ambient concentrations to ozone exposures of a study population depends on several factors, including the spatial distribution of ambient ozone and the typical activity patterns and home-ventilation patterns of the study community. Spatial variability in ambient ozone concentrations is generally less important in that concentrations of ambient ozone and other secondary pollutants tend to be relatively homogeneous (e.g., Wade et al. 2006) over large areas. Over greater distances, ambient ozone concentrations do vary, being sometimes higher in suburban and rural areas than in urban areas (Liu et al. 1993; Waldman et al. 1990) and higher away from roads than near roads. Transport of ozone and its precursors can also lead to a gradient in concentrations in an area.

Such causes of spatial variability in ozone concentrations in a community increase exposure error in time-series studies based on ambient concentrations, because this spatial variability can vary over time. The increased error may result in biased risk estimates (Wakefield and Shaddick 2006). Nevertheless, spatial variability in ozone concentrations is generally a minor concern in time-series epidemiologic studies because other factors, such as indoor concentrations and activity patterns, usually introduce larger errors in estimates of ozone exposure over time. Indoor ozone generally originates outdoors; indoor sources—such as photocopiers, laser printers, and some air cleaners—are not present in most homes. Indoor residential ozone concentrations are thus determined primarily by outdoor concentrations and by factors that affect the ability of ozone to penetrate into and persist in the home, including removal by reaction on surfaces, air exchange between indoors and outdoors, air filtration, and reactions between ozone and other indoor pollutants.

Under normal conditions, the half-life of ozone indoors is only 7-10 min (Weschler 2000). For averaging periods of 24 h, the short half-life results in generally low indoor concentrations (and often below the sampling method limit of detection).[2] These low concentrations reflect high rates of ozone removal by surface reactions. Indoor ozone concentrations are typically 10-50% of outdoor concentrations (Weschler 2006). Even at their lows, 24-h indoor ozone concentrations exhibit some diurnal and seasonal variation that reflects the outdoor variation. Indoor ozone concentrations tend to be higher in summer than in winter, when they are generally at or below the limit of detection of some measurement methods in personal exposure studies (Liu et al. 1993; 1997; Avol et al. 1998; Sarnat et al. 2006). For example, in a Southern California study of 126 homes (February-December 1994), mean 24-h indoor ozone concentrations (13 ± 12 ppb) were lower than corresponding outdoor concentrations (37 ± 19 ppb) (Liu et al. 1997). Indoor:outdoor ratios were greater during the summer pollution period. Higher summer ratios indicate the importance of home ventilation, especially in homes that do not have air conditioning or tight sealing. In non-air-conditioned and other well-ventilated homes, 24-h indoor concentrations can be a larger fraction of corresponding outdoor concentrations (Liu et al. 1993; Gold et al. 1996; Sarnat et al. 2006; Weschler 2006).

Similarly, in shorter averaging periods, indoor residential ozone concentrations can be a larger fraction of those outdoors when windows are opened. In

---

[2]With the exception of periods when indoor concentrations of nitric oxide (NO) are elevated, indoor ozone concentrations are normally not below the limit of detection of UV photometric methods (Weschler and Shields 1994).

a study by Zhang and Lioy (1994), for example, indoor ozone concentrations in six New Jersey homes measured during the high-ozone afternoon hours (2:30-7:30 pm) were found to track outdoor concentrations, with a mean ratio of 5-h indoor to outdoor ozone concentrations of 0.22 ($\pm$ 0.09) to 0.62 ($\pm$ 0.11). The indoor:outdoor ratios were highest (0.59 $\pm$ 0.16) when windows were open and gas stoves were off. For homes with air conditioning and homes with closed windows and no air conditioning the indoor:outdoor ratios over the 5-h high-ozone periods remained relatively low, 0.28 ($\pm$ 0.12) and 0.26 ($\pm$ 0.12), respectively. Use of a gas stove lowered indoor:outdoor ratios further for all ventilation conditions.

The low indoor ozone concentrations explain why 24-h personal ozone exposures are low: people spend on the average 87% of their time indoors (Klepeis et al. 2001). The average amount of time people spend outside during their day varies little across the United States (Klepeis et al. 2001). Therefore, daily averaged ozone concentrations to which people are exposed (averaged over the time spent indoors and outdoors) are generally low relative to outdoor concentrations averaged over the same period regardless of geographic location, with measured personal ozone concentrations generally below the limit of detection of some measurement methods. Like 24-h outdoor and indoor concentrations, 24-h personal ozone exposures are higher in summer than in winter, although summertime concentrations to which individuals are exposed (averaged over 24-h) are still a relatively small fraction of corresponding 24-h outdoor values (Liu et al. 1993, 1997; Sarnat et al. 2000; Sarnat et al. 2006). For example, in a Steubenville, OH, cohort of older adults which was conducted over a seven month period, the mean 24-h personal:ambient ratio of ozone concentration, 0.24, was substantially lower than that of any of the other measured pollutants ($PM_{2.5}$, EC, sulfate, NO_2, and SO_2).

It is important to note that over periods of less than a day, such as 1 h, personal ozone exposures can be relatively high; Hourly personal exposure concentrations were shown in a scripted-exposure study to be comparable with corresponding outdoor concentrations when the trained technician spent the time outdoors (Chang et al. 2000). Based on these findings, it is possible that hourly and peak ambient ozone concentrations are appropriate surrogates for personal ozone exposure concentration, especially in summer and during peak afternoon high ozone hours, which account for much of the individual outdoor time (Figure 3-9). Further study should examine whether this is in fact the case, especially for individuals who are susceptible to ozone exposures.

Hourly personal exposures in the scripted-exposure study were also strongly associated with hourly outdoor concentrations when the trained technician was outdoors (Chang et al. 2000). In contrast, when the technician spent the hour indoors or in transit, associations between hourly personal exposures and ambient concentrations were not significant. Similarly, over 24 h, the vast majority of which is spent indoors, many studies suggest that personal ozone exposure concentrations are not strongly associated with corresponding ambient con-

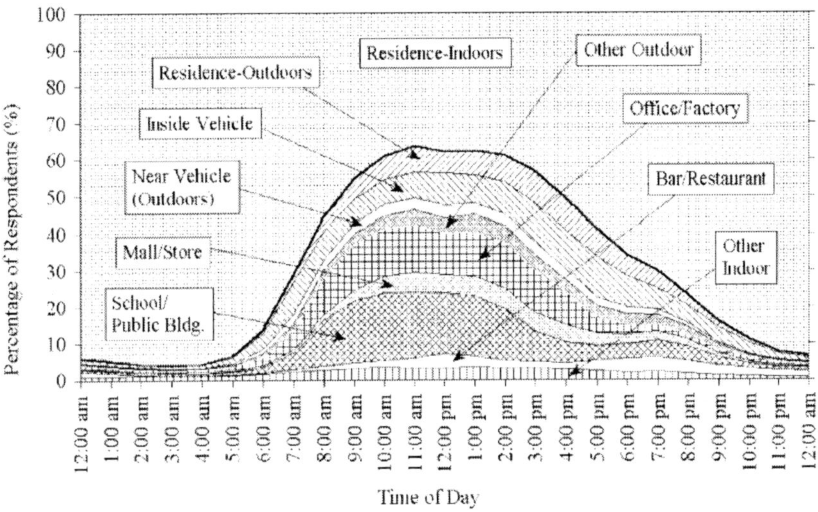

FIGURE 3-9 Activity patterns by time of day. Adapted from U.S. National Human Activity Pattern Survey. Source: Klepeis et al. 2001. Reprinted with permission; copyright 2001, *Journal of Exposure Science and Environmental Epidemiology*.

centrations regardless of whether individuals are followed over time or at one point in time, with consistently low slopes and R^2 values for regressions of personal ozone on ambient concentration (Brauer et al. 1989; Liu et al. 1993, 1997; Linaker et al. 2000; Patterson and Eatough 2000; Sarnat et al. 2000; Sarnat et al. 2006). It is important to note that these low R^2 values may be due in part to the low measured personal exposures, which were often below the limit of detection of the measurement methods used in those studies. Despite this, the low R^2 values do suggest that estimates from time-series studies may be biased toward the null hypothesis.

More recent studies suggest that the associations are substantially weaker than those for $PM_{2.5}$ and sulfate (Sarnat et al. 2000; Sarnat et al. 2006) and stronger in summer than in winter. The seasonal variability probably reflects increased home ventilation in the hotter summer months. Consistently with that, the slope of the regression between 24-h ambient ozone concentrations and corresponding personal exposures in Steubenville, OH, was 100% higher for elderly people spending time in well-ventilated indoor environments (slope, 0.18 ± 0.03; t value, 7.34) than for other people (slope, 0.08 ± 0.04; t value, 1.89). As supported by Levy et al. (2005), those findings suggest that ventilation (and thus season) is an important modifier of ozone-mortality risk estimates, and further provide potential explanations for observed seasonal or between study variability in other studies.

Even in well-ventilated conditions, however, the slope of the ozone association is small, and this suggests that only minor changes in ozone exposure occur in response to moderate changes in outdoor concentration. Together, low personal exposure concentrations and weak personal-ambient concentration relationships suggest that 24-h ambient ozone concentrations are poor proxies for personal exposure. For shorter averaging periods, results of the one scripted-exposure study suggest that hourly or peak ambient ozone concentrations may be appropriate proxies for corresponding hourly or peak personal exposures. Additional short-term personal-ozone exposure studies are needed, given ozone's large contribution to uncertainty in ozone-mortality risk estimates; however, these studies will require the development of new measurement methods that have sufficient sensitivity to measure these likely low, short-term exposures.

The relatively poor correlation of ambient ozone concentration with personal exposure—combined with the strong associations among ambient ozone concentration, ambient $PM_{2.5}$ concentration, and temperature—raises concerns about whether ozone mortality attributed to ozone is due instead to these or other correlated factors. Evidence from a series of 24-h multipollutant exposure panel studies performed generally over two seasons at locations—including Baltimore, MD (Sarnat et al. 2001), Steubenville, OH (Sarnat et al. 2005), and Boston, MA (Rojas-Bracho et al. 2004)—indicates that potential confounding by correlated pollutants and weather is an important source of uncertainty in ozone-mortality risk estimates. Results from longitudinal analyses have consistently shown, for example, that 24-h ambient ozone concentrations are substantially weaker proxies for corresponding exposures than $PM_{2.5}$. In addition, studies have shown 24-h ambient ozone concentrations to be important proxies for personal $PM_{2.5}$ exposures over time. Associations between 24-h ambient ozone concentrations and personal $PM_{2.5}$ exposures tended to be stronger in Boston and Steubenville than in Baltimore, possibly because of better home ventilation in the two cities. It is consistent with that idea that for all cities 24-h ambient ozone was a stronger proxy for personal $PM_{2.5}$ exposures for individuals followed in summer than winter and for people spending most of their time in well-ventilated than in poorly ventilated environments. Because exposure studies of confounding have been based on 24-h values and for individuals living within the eastern United States, the generalizability of these results to exposures measured over shorter periods and to other populations is not known.

OZONE-EXPOSURE MODELING

One method for linking personal ozone exposure to ambient ozone concentration, and thus linking ambient ozone to health end points, is to model exposure in ways that account for individual behavior and individual ozone exposure concentrations. Such simulation models are valuable for estimating air-pollution exposures in the absence of direct measurement. They can provide a framework for extending risk analyses across large geographic domains, even at the national level. They can also be used to help to estimate the effect of ambient pollutant reductions on personal exposure, as is done in developing the ozone criteria document. Such analysis has been used in advising the EPA administrator on proposing standards (e.g., EPA 2007a).

Exposure-simulation models can be used in epidemiologic models to study the association between human exposure to ozone and mortality or morbidity instead of relying on a monitored ozone value and providing a potentially more accurate representation of individual ozone exposure. Such human-exposure models could be powerful tools for drawing inferences about the health risks associated with exposure to tropospheric ozone. That approach was taken in the recent EPA ozone staff paper as part of the policy assessment (EPA 2007a). Uncertainty in such models should be characterized and quantified properly before they are used for ozone-mortality risk assessment for regulatory purposes.

The ozone-exposure model developed by EPA to estimate human population exposure to the criteria and other toxic pollutants is called Air Pollutants Exposure (APEX). APEX is a probabilistic model designed to account for the numerous sources of variability that affect people's exposures. APEX simulates the movement of people through time and space and estimates their exposure to a given pollutant in indoor, outdoor, and vehicular microenvironments. Daily activity patterns for individuals in a study area, one of the inputs to APEX, are estimated from detailed time-location-activity diaries that are compiled in the Consolidated Human Activity Database (CHAD) (McCurdy et al. 2000; EPA 2003d). The diaries contain information regarding age, sex, race, employment status, occupation, day of week, daily maximum hourly average temperature, each location during the day, and start time, duration, and type of activity performed. APEX estimates the concentration in the microenvironment associated with each event in an individual's activity pattern and sums the event-specific exposures within each hour to obtain a continuous series of hourly exposures spanning the period of interest. Activity-specific simulated breathing rates of individuals are used in APEX to characterize intake during each period. The breathing, or ventilation, rates are derived from energy-expenditure estimates for each activity included in CHAD and are adjusted for age-specific and sex-specific physiologic characteristics associated with each simulated individual.

Ozone concentrations in each microenvironment are estimated with mass-balance or transfer factors, and the user specifies prior probability distributions for the parameters to be used in the model; the prior distributions are used to model the uncertainties and variabilities in the parameters.

APEX combines the estimated time series of exposure concentrations that a simulated individual experiences during the modeled period with the estimated time spent in each of the microenvironments visited according to the activity diary. The hourly average exposures of each simulated individual are time-weighted averages of the within-hour exposures. APEX then statistically summarizes and tabulates the 1-h, 8-h, and 24-h individual exposures, and then provides a distribution of ozone exposure for the population in a given census as characterized by the simulated individuals. The estimated exposure is a linear function of the ambient ozone concentrations.

Although there is the potential to use those exposure simulation models for future analyses, APEX has not yet been used to characterize ozone exposure in epidemiologic models for national ozone mortality analysis. Therefore, it has not been given more extensive consideration in this report.

The parameter uncertainty in APEX is represented by prior distributions assigned to most of the parameters in the model. The prior distributions given to the model parameters are based on regional studies; to extend the prior distributions to the national level, more exposure studies should be conducted. There also is uncertainty in the formulation of the model and in the inputs. The different sources of error and uncertainties in APEX result from variability that is not modeled or is modeled incorrectly; from erroneous or uncertain inputs; from errors in coding; from simplifications of physical, chemical, and biologic processes to form the conceptual models; and from flaws in the conceptual model. In particular, the uncertainty in the estimation of ambient air quality will be propagated by APEX. The APEX output may also be very sensitive to the prior distributions used in the microenvironmental models. APEX relies heavily on the assumption that exposure to ozone is a linear function of ambient concentration, and this assumption needs more justification.

CHAD may be the best available source of human-activity data for use in exposure modeling, but issues regarding how well CHAD diaries represent the simulated populations remain. CHAD contains 20,000 people that are used to represent several million over long periods, the diary data are relatively old (some data were generated in the 1980s), and there are diary structure differences (real-time data collection vs data collection by recall); each of these characteristics may lead to errors. The human diaries used by the model might be too limited to characterize and represent the simulated populations. EPA has conducted some validation studies of APEX (EPA 2007k). The committee agrees that the APEX approach is a useful way to assess population exposure, but remaining uncertainties require further study to estimate their magnitude and, to the extent possible, reduce them.

SUMMARY

Interpretation of studies relating measurements of ambient ozone to health end points is made difficult by the dynamics of ozone and associated pollutants, particularly how temporal and spatial variations affect individual human exposure. (Other questions about the existing evidence are addressed in Chapter 4.) Correlations between ozone and $PM_{2.5}$ vary spatially, seasonally, and with $PM_{2.5}$ components, all of which can confound the interpretation of epidemiologic analyses. Furthermore, exposure studies which typically relied on passive ozone monitors found that ambient ozone concentrations are not highly correlated with concentrations indoors, where people spend most of their lives, or with observed personal exposures. Indoor ozone concentrations exhibit some diurnal and seasonal variations that reflect the outdoor variations. Correlations with ambient concentrations may be artificially low given that measured 24-h personal exposures concentrations and indoor concentrations were often below the method limit of detection. One way to improve characterization of indoor ozone exposure is to use continuous ozone monitors, which have greater sensitivity than passive ozone monitors. At the population level, characterization of ozone exposures can be improved using exposure modeling, although additional uncertainties are involved in model application. Accounting for those complexities presents a formidable obstacle in interpreting results of epidemiologic studies of ozone and health.

Finding: Potential confounding of ozone-mortality studies by PM presents a challenge, particularly in considering seasonal and component effects. Some species are much more (or less) correlated with ozone when specific seasons are considered (for example, summer, when ozone is high) than over an annual cycle. Considering only PM mass would not account for effects of specific PM components.

Recommendation: Both exposure and epidemiologic researchers should investigate seasonal and regional associations between ambient concentrations and exposures to ozone and $PM_{2.5}$, (and its components), how they affect control of confounding, and correlations between the various pollutant concentrations. When possible, researchers should address those issues by focusing on groups of individuals who are sensitive to ozone exposures and by using data on the chemical and physical components and size distribution of $PM_{2.5}$.

EPA and the scientific community should increasingly include the growing Speciated Trends Network (STN) database in future analyses of potential confounding of the ozone associations. EPA should work with the scientific community to ensure that the STN collects data frequently enough on the particle components most relevant to the potential for confounding.

Finding: Personal exposure, considering time spent outdoors and indoors on a 24-h basis, is poorly correlated with monitored ambient ozone concentrations, with low slopes of the regression of personal exposure on ambient ozone levels. However, findings on exposure in the afternoon, when both outdoor activity and ozone concentrations can peak, suggest that exposures to higher ozone concentrations are better captured by ambient monitoring. Control programs based on reducing peak afternoon ozone concentrations on days when they are greater than the standard can result in an increase in the 24-h concentrations on days that are conducive to ozone formation and can also increase afternoon concentrations on days less conducive to ozone formation.

Recommendation: Future studies should determine whether and how much daily peak exposures, such as 1h or 8h exposures, and longer-term average exposures, such as over 24h, are associated with ozone-related mortality so as to guide control decisions for protecting public health. Benefits assessors at EPA and elsewhere should use the results to identify the appropriate exposure metrics to estimate how efforts to attain the ozone NAAQS will affect ozone exposure and health.

The committee notes that EPA has apparently not considered the use of two or more averaging times jointly. Does adding metric C to a model that already contains metric A, or metrics A and B, improve the fit of the model? Such analyses should be fairly easy to conduct and might show that 1-hour, 8-hour, or 24-hour averaging times are roughly equivalent and that any one of these can substitute for the others. Conversely, one might find that one or two of these metrics add little when the third is already in the model. Or, it is possible that some combination of two or three of these is a better measure for analysis than any one alone. Measurement error is likely to blur the actual contributions of each metric and the relationships among them, but that is not relevant when the goal is to find the best empirical fit to the data.

Regulators need to consider that control strategies may affect 24-h average concentrations quite differently from how they affect shorter-term exposures and that peak short-term concentrations on lower-ozone days will respond differently, and often in the opposite direction, from the same measure of ozone on a high-ozone day.

Finding: Human-exposure simulation models, such as APEX, present an approach to estimating ozone exposure different from reliance solely on observations. However, such models introduce their own uncertainties, and they need to be further evaluated and their uncertainties characterized.

Recommendation: EPA should conduct more detailed evaluations of APEX and other models used to improve characterization of ozone exposure at the population level by taking human activity into account. Detailed evaluation will require further human-exposure studies, which will need improved instrumentation and approaches for monitoring personal exposures. The extent to which the human diaries represent the population under study, as used by APEX, warrants further explication, and more extensive and more up-to-date diary data should be collected, in particular for children and minority groups. Those data, when tied to data on pollution concentrations and ozone-alert days, can be used to assess the type and degree of changes in behavior to avoid ozone exposure. The cost of such changes in behavior associated with increased pollution currently is not addressed in RIAs prepared by EPA.

Finding: Epidemiologic studies of ozone health effects are limited, in part, by the reduced availability of ozone data in winter.

Recommendation: EPA and states should extend operation of ozone monitoring into winter and report the measurements. The size of the winter program should be sufficient to enhance researchers' ability to examine (1) seasonal differences in risk, (2) how these seasonal risk differences vary spatially between communities with warmer and cooler winters, and (3) ozone-mortality relationships at lower ozone concentrations. In recognition that ozone is a regional pollutant, winter measurements need not be collected in all the summer locations but, when they are collected, should be collected at the frequency of summer measurements.

4

Contributions of Relevant Health Studies to the Estimation of Reductions in Premature Mortality

INTRODUCTION

This chapter reviews evidence of ozone-mediated effects on mortality. It builds on the wealth of toxicologic and epidemiologic evidence showing the deleterious effects of ozone on an array of health effects, including oxidative damage to DNA, inflammation, oxidative stress, heart-rate variability, decrease in lung function, such respiratory conditions as bronchitis and wheezing, increase in medication use, and exacerbation of respiratory diseases leading to emergency-room visits and hospitalization. Those effects have consistently been found in toxicologic and epidemiologic studies, independently of study design and location, and therefore constitute compelling and well-accepted evidence that ozone adversely affects health.

In contrast, until its regulatory impacts assessment (RIA) for the proposed ozone national ambient air quality standards (NAAQS) released in 2007, the Environmental Protection Agency (EPA) had not included mortality results in it primary estimates of the benefits of reductions in acute exposure to ambient ozone, because of uncertainty about the appropriate interpretation of those results. Previously, EPA had included epidemiologic studies that link ambient ozone concentrations with premature deaths in sensitivity analyses. The shift is attributable to consistent findings from multicity time-series studies and meta-analyses conducted over the past few years that suggested a modest but highly significant increase in mortality associated with relatively short-term exposure to ambient ozone.

This chapter reviews evidence of ozone-related death with regard to its biologic plausibility and reviews findings from the multicity time-series studies and meta-analyses. It also assesses evidence of the existence of susceptible groups and discusses how such factors as personal and population average-exposure error, epidemiologic designs and methods of analysis, measures of the effects of ozone exposure on death, and uncertainty contribute to our ability to estimate ozone-related mortality risk and to apply the estimates to risk and benefits analysis.

BIOLOGIC PLAUSIBILITY

The mechanisms by which ozone can damage health and possibly lead to mortality remain a subject of considerable clinical interest. Toxicologic studies have attempted to identify the mechanisms by using a variety of techniques, from ultrastructural, biochemical, and cytologic analyses to in vivo measurement of airflow mechanics. The techniques generally examine the effects of controlled ozone exposures of less than 8 h and, to a smaller extent, exposures of several days to increased ambient concentrations.

The findings of a number of toxicologic studies provide insights to suggest the potential for a number of events by which ozone exposure could lead to increased mortality, including lung inflammation leading to local pulmonary compromise (Devlin et al. 1991; Holz et al. 1999; Koren et al. 1991; Ratto et al. 2006; Bosson et al. 2007), worsening of pre-existing cardiopulmonary disease, systemic release of mediators that affect adverse cardiovascular events (Hollingsworth et al. 2007), effects on the autonomic nervous system that could contribute to increased airway responsiveness (Chen et al. 2003) or reduced heart rate variability, and an increase in factors that lead to vascular changes (Chuang et al. 2007). Definitive proof of such occurrences leading to mortality are not found in the current scientific literature, however, there is ample evidence ozone can induce mechanisms through a sequence of events to trigger oxidative stress or inflammation pathways, which can contribute to release of pulmonary or systemic mediators. The events may further result in the exacerbation of a pre-existing respiratory or cardiovascular condition. Chen et al. (2004) found increased sensitivity to aeroallergens in a subgroup of asthmatics following exposure to 200 ppb ozone. There exists the biological plausibility that for a few individuals with pre-existing cardiopulmonary or chronic respiratory disease, such oxidative or inflammatory events could also cause death, via the onset of an asthmatic attack (Selgrade et al. 2008), myocardial infarction, or vascular event. Inflammatory events have been shown to occur for 20-30 h after inhalation of ozone even at ambient concentrations in the range of 0.08-0.10 ppm (Ratto et al. 2006; Chuang et al. 2007). There is less information on the airway inflammatory response to ozone at low concentrations, such as at or below 0.08 ppm.

The oxidative stress and inflammation cascade is initiated by an ozone-mediated airway and lung tissue response (Chuang et al. 2007). Ozone can initiate the cascade immediately after inhalation when it comes in contact with epithelial lining fluids and cellular membranes, most often at nasal epithelial surfaces and at the junction of the bronchioles and the alveolar region (centriacinar region). At the air-liquid interfaces, ozone can interact with unsaturated fatty acids to form lipid ozonation products, which in turn can activate lipases that lead to production and release of cell-signal transduction molecules and proinflammatory mediators. With oxygen species and reactive oxygen metabolites generated by exposure, those products of ozone exposure may lead to oxidative stress and the upregulation of transcription factors, such as NF-κB and proinflammatory genes. Correspondingly, antioxidant enzymes (such as glutathione enzymes, superoxide dismutase, and catalase) and nonenzymatic factors (such as ascorbic acid and α-tocopherol) that react with ozone directly have been shown to attenuate oxidant damage to airway lipid membranes.

Biochemical evidence of ozone-mediated inflammation is provided by animal and human studies that show increases in downstream activation products of the innate immune system, including fibronectin, elastase, plasminogen activator, tissue factor, factor VIII, C3a fragment of complement, prostaglandins, interleukin-1 (IL-1), tumor-necrosis factor a, IL-6, IL-8, and granulocyte macrophage colony-stimulating factor (GM-CSF) in alveolar lavage fluid (Koren et al. 1989; Aris et al. 1993; Devlin et al. 1994). Inflammatory effects of ozone occur in people who are healthy and in people who have increased vulnerability because of host (genetic) factors or pre-existing cardiopulmonary disease, such as asthma. For example, ozone exposure at 100 ppb for 2 h leads to an acute neutrophilic inflammation of the airway in allergic asthmatic subjects (Depuydt et al. 2002). In addition, pre-exposure of these subjects (atopic asthmatics) to ozone potentiates the late-phase eosinophilic response to the allergen. The acute lung response to ambient ozone has many clinical features in common with asthma, and ozone exposure also leads to increase in proinflammatory cytokines (Koren et al.1989), cellular inflammation in airway tissues, and bronchial hyperresponsiveness (Foster et al. 2000). Perhaps as a result, exposure to ambient ozone can exacerbate pre-existing allergic asthma (White et al. 1994). Cultured human epithelial cells (HBECs) from asthmatic and nonasthmatic subjects exposed for 6 h to ozone at 100 ppb had increases in inflammation-associated mediators IL-8, GM-CSF, and intercellular adhesion molecule 1 (Bayram et al. 2001). HBECs from asthmatic but not nonasthmatic people also showed increased expression of the chemotactic cytokine CCL5. Increased expression of CCL5 suggests that ozone exposure may increase eosinophilic inflammation, exacerbate asthmatic symptoms, and lead to pathologic changes in the epithelium (Bayram et al. 2001). Support for that suggestion is provided by studies that show human asthmatic subjects to have increased eosinophilic inflammation of the lower airways after prolonged acute exposure to ozone at 160 ppb. Epithelial cytokine expression of IL-5, GM-CSF, and IL-8 has also been found to be higher in asthmatic subjects after laboratory exposure to ozone for 2 h at 200 ppb (Bosson et al. 2003).

Numerous laboratory chamber studies have validated the reproducible nature of the inflammatory lung response (Devlin et al. 1991; Koren et al. 1991; Holz et al. 1999). That finding has been supported by additional studies in humans using therapeutic ablation of inflammatory effects, which have also helped uncover essential mechanisms of injury. Although inhalation of the corticosteroid budesonide was not found to limit ozone-induced airway inflammation in healthy subjects, budesonide reduced neutrophilic infiltration in patients with mild asthma as evidenced by changes in the inflammatory cellularity of induced sputum (Nightingale et al. 2000; Vagaggini et al. 2001). Inhaled and oral corticosteroids have also been found to ablate ozone-induced neutrophilic inflammation in selected healthy subjects who were characterized as neutrophilic ozone-responsive (Holz et al. 2005). In mildly asthmatic subjects, pretreatment and posttreatment with apocynin aerosol, an inhibitor of the nictotinamide adenine dinucleotide phosphate (NADPH) oxidase complex present in membranes of inflammatory cells (eosinophils and neutrophils) and a source of reactive oxygen species, prevented ozone-induced airway hyperresponsiveness to methacholine (Peters et al. 2001).

There has been relatively little research done on cardiovascular effects of ozone as compared with the amount done on pulmonary effects. Gong et al. (1998) used an ozone chamber study to monitored cardio-vascular effects in humans. In response to exposures to 300 ppb ozone, no major acute cardiovascular changes were found to be induced in either hypertensive or control subjects. It was noted, however, that ozone can increase myocardial work and impair pulmonary gas exchange.

Together, those findings constitute supporting evidence that ozone exposure induces a cascade of events that increase oxidative stress and inflammation. As observed in numerous cohort and toxicologic studies, it is plausible that increased oxidative stress and inflammation can influence cardiac risk and ultimately mortality through increased autonomic dysregulation, vascular dysfunction, atherosclerosis, and arrhythmogenesis (Ridker et al. 1998, 2000; Pradhan et al. 2001, 2002). Those downstream effects may be most pronounced in people who have pre-existing diseases, in that ozone exposure may worsen a disturbed respiratory, vascular, or cardiac system, making them more vulnerable to ozone's adverse effects.

The adverse effects can occur after intermittent exposure to ozone, which is a typical pattern of exposure to this pollutant. Although the effects of ozone may accumulate over a lifetime, it is well established that gaps in ozone exposure lead to a decrease in innate resistance to its adverse effects. With the onset of each ozone-exposure episode, the injury process will be repeated, initiating once again the biologic events listed above. Such a cellular response to ozone means that each exposure event poses a renewed risk of a cascade of activities that could decrease resiliency and possibly enhance vulnerability, resulting in premature death.

OZONE-MORTALITY STUDIES

Given the evidence from toxicologic studies, it is biologically plausible that ozone exposures are also related to death in human populations. From a biologic and epidemiologic perspective, there are two main ways in which air pollutants, such as ozone, may affect mortality: as a consequence of acute effects that cause death in the near term and as a consequence of chronic pathophysiologic changes that ultimately lead to death. The two may well overlap and be interrelated, but the distinction is useful because it is relevant to the interpretation and quantification of risks.

Acute Studies

Acute effects of ozone are those observed within a few hours or days of a rather short exposure. In those cases, exposure is typically defined as the maximum concentration on a given day, the mean across the 8 consecutive hours with the highest concentrations, or the 24-h average concentration. Under this acute-effect model, if death occurs, it is a consequence of the exposure. An acute exposure may also trigger a condition of poor health or frailty that can lead to death within a few days or weeks if intervention is not successful; in this delayed case, death may be considered a subacute effect of an acute (short-term) exposure.

Time-series analysis is by far the most common approach to investigate the acute and subacute effects of ozone on mortality (see Box 4-1). The most systematic and comprehensive time-series analyses were two multicity studies (Bell et al. 2004; Ito et al. 2005) and three meta-analyses (Bell et al. 2005; Ito et al. 2005; Levy et al. 2005) solicited and commissioned by EPA.[1] The three research teams conducting the meta-analyses were provided with the same databases from EPA but conducted the analyses separately and did not communicate with each other about their methods or findings until the studies were completed. Those studies examined ozone-mediated mortality while addressing previously unresolved issues related to confounding, exposure variability, and model specifications. Although the studies used different approaches, their major results were similar: each found a statistically significant relationship between ozone exposure and premature mortality that appears robust after controlling for exposure to particulate matter with a diameter of 10 µm or less (PM_{10}). The similar estimates from these three meta-analyses suggest not only an indication of stable estimates in the literature, but also indicate that the meta-analytic approach is generally insensitive to analytical decisions, which provides support for the meta-analytical methods. Together, their results were cited by EPA as constituting strong evidence of a link between short-term exposure at concentrations below the NAAQS and premature mortality. However, many questions remain, specifically regarding the size and significance of the ozone-mortality effect and the implications of the findings for ozone benefits analysis. Study

---

[1]The study by Ito et al. (2005) included both a meta-analysis and a time-series analysis; results from the time-series analysis were intended to help to explain issues identified in the meta-analysis.

BOX 4-1 Definitions of Time-Series Analysis and Meta-Analysis

Time-series analysis describes and models the behavior of observations that are occurring sequentially and is analysis of the temporal relation between ozone and mortality data at different times. Moreover, time series analysis in air pollution studies is typically an approach in which daily mortality or morbidity counts are expressed as a function of covariates, including at least one indicator of air pollution in regression models, which can control for confounding effects of seasonality, long term trend and weather. The multicity time-series model has several advantages over multicity meta-analyses. Instead of having to rely on the different analytic approaches and assumptions used by researchers of completed studies, the multi-city time-series model provides the ability to use an identical model structure, exposure lags, control for confounders, and statistical criteria for data from all cities. Furthermore, the time-series model can include factors that explain differences in effect estimates among the cities, such as region, housing characteristics, and measurement method.

Meta-analysis is a statistical technique used to aggregate, summarize, and review previous quantitative research. Through meta-analysis, a wide variety of questions can be investigated, assuming that a representative body of primary research exists. Selected parts of the reported results of primary studies (effect sizes or other characteristics) are analyzed as other data are: descriptively and then inferentially to test hypotheses. The appeal of meta-analysis is that it in effect combines research on a topic into one large study with many subjects. The danger is that in aggregating a large set of studies, the construct definitions can become imprecise and the results difficult to interpret fully (Neill 2006). Moreover, a major issue with meta-analysis is publication bias, which refers to the greater likelihood that studies with statistically significant positive results will be published as compared with studies reporting negative results. Thus, estimates from pooled meta-analysis may overestimate the association between ozone and mortality.

Like multicity time-series analyses, meta-analyses use a multilevel model to examine pollution-effect associations. City specific estimates (level 2) are obtained from time series data, (level 1) for each city with a generalized linear model analysis. These are then pooled into a national average estimate using a Bayesian hierarchical model in which stages of the hierarchy have assumed normal distributions.

descriptions, findings, limitations, and remaining issues related to the five comprehensive analyses are presented below.

Multicity Time-Series Studies

Two multicity time-series studies of ozone and mortality were conducted. Bell et al. (2004) used the National Morbidity, Mortality, and Air Pollution

Study (NMMAPS) database to conduct a large time-series analysis with data from 95 U.S. cities over the period 1987-2000. Ito et al. 2005 conducted a time-series study with data from seven U.S. metropolitan areas over the period 1985-1995.

Bell et al. (2004). As shown in Table 4-1, Bell et al. used data and methods developed for the NMMAPS study to estimate a national average relative rate of mortality (noninjury mortality and cardiovascular and respiratory mortality) associated with short-term average ambient ozone concentrations in 1987-2000 in 95 large U.S. urban communities made up of almost 40% of the U.S. population. The 95 areas were chosen on the basis of availability of daily ozone data from EPA's Aerometric Information Retrieval Service (now known as the Air Quality System database). Descriptive statistics on each community are provided at (iHAPSS 2005). For ozone, the 24-h average, maximum 8-h, and maximum hourly concentrations were calculated for each day. In several locations, ozone was measured only during the peak ozone season, April-October. A two-stage statistical model was used to estimate a national average association between short-term ambient ozone concentrations and mortality risks, accounting for weather, seasonality, long-term trends, and PM_{10}. In the first stage, distributed-lag overdispersed Poisson regression models were used for estimating community-specific relative rates of mortality associated with exposure to ozone in the week prior to death. In the second stage, the community-specific relative rates were combined to generate a national average estimate of the association between ozone and mortality that accounts for within-community and across-community variability.

Analyses were performed with multiple ozone-concentration metrics (1-h maximum, 8-h maximum, and 24-h average), which were highly correlated with each other; variations were most likely due to weather and sources of ozone. Temperature was modeled as a natural cubic spline (a type of mathematical function) of a day's temperature and the average of the three previous days' temperatures. Analyses were based on single-day lags and constrained and unconstrained distributed-lag models for all 95 communities, for all communities in warm months (April-October), and for the 55 communities on which yearly ozone data were available. The sensitivity of model results to choice of model, temperature, age (all ages and less than 65, 65-74, and over 75 years), and PM_{10} was also assessed. The sensitivity analysis with PM_{10} was limited to days for which PM_{10} and ozone data were available (1 in 6 days).

Results showed that average short-term changes in ozone are significantly associated with premature mortality and that the statistical association is robust to adjustments for PM_{10}, weather (temperature), and seasonality. It should be noted that robustness for PM_{10} may have been hindered by lack of daily monitoring data. Ozone-mediated risks were greater for cardiovascular and respiratory mortality than for total mortality, and effects were larger at lag 0 d than lag 1 or 2 d. Bell et al. examined the 95 communities using a constrained distributed-lag model and found a statistically significant 0.52% (95% posterior

TABLE 4-1 Summaries of Recent Studies of Acute Effects of Ozone on Mortality

Author	Source	Design/ Number of Studies or Cities	Period	Exposure Metrics	Health Outcomes	Risk (95% Confidence Interval) per 10 ppb
Bell, McDermott, Zeger, Samet, and Dominici al.	JAMA 2004; 292(19): 2372-2378	Time-series study of 95 U.S. large urban areas	1987-2000	Cumulative exposure of previous week	Noninjury mortality	0.52% (0.27-0.77%)
					Cardiovascular and respiratory mortality	0.64% (0.31-0.98%)
Ito, Leon, and Lippmann	Epidemiology 2005;16(4): 446-457	Meta-analysis of 43 studies (international)	1990-2003	24-h average	Nonaccidental mortality	0.8% (0.55 -1.0%)
		Time-series analysis of 7 U.S. cities	1985-1995	0- and 1-d lag, 24-h average		0.52-1.0%[a]
Levy, Chemerynski, and Sarnat	Epidemiology 2005;16(4): 458-468	Meta-regression of 28 time-series studies (international)	Pre-October 2003	1-h maximum	All-causes mortality	0.41% (0.32-0.52%)
				1-h maximum: summer;		0.84% (0.57-1.09%)
				1-h maximum: winter		-0.04% (-0.34 to 0.28%)
Bell, Dominici, and Samet	Epidemiology 2005;16(4): 436-445	Meta-analysis of 39 time-series studies	1990-June 2004	0-, 1-, and 2-d lag or 2-d average of lags (0, 1, and 2 d)	Total mortality	0.87% (0.55-1.18%)

[a]Range of risk estimates without confidence intervals.

interval: 0.27%-0.77%) increase in non-accidental mortality for a 10 ppb increase in the previous week's daily average ozone concentration. The increase in cardiovascular and respiratory mortality was somewhat higher at 0.64% (95% PI, 0.31%-0.98%). Interestingly, Bell et al. found a lower increase in total mortality of 0.39% (95% posterior interval: 0.13%-0.65%) in those communities during the ozone high season (April through October). Results were similar for the unconstrained models.

The mortality effect of a single day of ozone exposure appeared to be distributed over several days. The study further suggested that ozone-mortality effects can be calculated separately from PM-mortality effects because ozone risk estimates were robust to adjustment for PM_{10}. Bell et al. (2007) also reported the 95 Bayesian city-specific estimates of risk and the standard deviation of these city-specific estimates. The standard deviation for the constrained distributed lag model was 0.64%. Thus 95% of the true city-specific risk effects lie in the interval -0.73% to 1.77%. Bell at al. (2007) further investigated whether PM is a confounder of ozone and mortality using data for 98 U.S. urban communities from 1987 to 2000. They concluded that neither PM_{10} nor $PM_{2.5}$ is a likely confounder of observed ozone and mortality relationships. They recommended that further investigation is needed of potential confounding of the short-term effects of ozone on mortality by PM chemical composition. As part of a larger study of ozone and mortality, Ito et al. (2005) conducted a time-series analysis with data from seven U.S. cities (Chicago, Detroit, Houston, Minneapolis-St Paul, New York City, Philadelphia, and St Louis) for the period 1985-1994 (Table 4-2). Those cities were chosen because of the availability of data on daily (or nearly daily) PM_{10}, which went back to 1985 in most cities, and their relatively large populations. It was necessary to obtain daily rather than every-6th-day samples (the usual sampling frequency for most U.S. cities) of PM to obtain reasonable statistical power (Ito, personal communication, 12/04/07). In each of the cities, ambient ozone was measured year-round, and daily PM_{10} data were available for all cities except New York. $PM_{2.5}$ data were available for selected years for Philadelphia (1992-1995) and St. Louis (1985-1989). Ito et al. (2005) used the data specifically to examine the sensitivity of ozone-mortality risk estimates to several factors, including season, alternative weather models and adjustments (of which four were examined), and confounding by PM. Results of the sensitivity analyses were intended to supplement a corresponding meta-analysis.

To characterize the ozone-PM relationships across seasons, the authors computed the mean ozone concentration for each quintile of PM in summer and winter. To examine the sensitivity of ozone-mortality risk estimates to alternative weather-model specifications and temporal adjustments, they used a Poisson generalized linear model that adjusted for temporal trends, day of week, and weather effects. The average of 0- to 1-d lagged 24-h average ozone concentration was included. A smoothing function of days using natural splines was included to adjust for seasonal cycles and other temporal trends, and various weather models from the literature were investigated.

TABLE 4-2 Summaries of Recent Time Series Studies of Acute Effects of Ozone on Mortality

Authors	Bell, McDermott, Zeger, Samet, and Dominici.	Ito, Leon, and Lippmann
Study design (methods)	Time-series study	Time-series study
Type of model(s)	Hierarchic (mixed-effects), distributed Lag	Poisson generalized linear model
Numbers of studies or cities	95 U.S. large urban areas	7 U.S. large urban areas
Study periods	1987-2000	1985-1995
Exposure level	County level	Metropolitan areas
Exposure metric	24-h average 1. Maximum 8-h average 2. Maximum hourly concentration	24-h average
Time of study	1. April-October 2. Whole year	Whole year
Health outcomes	1. Non-injury-related mortality 2. Cardiovascular and respiratory mortality	Total nonaccidental mortality
Lag time days in days	0, 1, 2, 3, up to a week	
Adjustment for temperature	Yes	Yes
Adjustment for humidity	No	No
Adjustment for season	Yes	Yes
Consideration of age strata	Yes	Yes
Adjustment for PM_{10}	Yes	Yes
Adjustment for $PM_{2.5}$	Yes	Yes
Adjustment for other pollutants	No	No
RESULTS		
Non-injury-related mortality (per 10-ppb change) 1-h maximum 8-h maximum 24-h average	0.39% (95% posterior interval, 0.13-0.65%) with up to a week lag (April-October)	1.0% (0 .55-1.40%) with 0-, 1-d lag.(year-round)
Cardiovascular and respiratory mortality (per 10-ppb change) 1-h maximum 8-h maximum 24-h average	0.64% (0.31-0.98%) with up to a week lag	

Results from the multicity analysis show short-term associations between ozone and mortality with substantial heterogeneity across cities. Pooled analysis for the six cities with corresponding daily PM data resulted in an all-year, ozone-only mortality effect estimate of 2.0% (1.1-2.9%) for the quintile model and 1.0% (0.0-2.0%) for the model that used four smoothing terms per 20-ppb increase in the average of 0- and 1-d lag 24-h ozone. When stratified by season, the excess risk estimates were higher in summer than in the whole year and in cold seasons, when the estimates were low or null. The potential confounding between ozone and PM did not substantially affect ozone risk estimates.

Results further showed that the use of different weather models could account for a twofold difference in overall effects estimates (0.24-0.49%) in analyses that included yearly data, with the quintile temperature model producing the largest estimate and the model that used four smoothing terms producing the smallest. Regardless of the adjustment for weather and for PM_{10}, $PM_{2.5}$, and season, however, large city-to-city variation in ozone-associated mortality risk estimates persisted. The heterogeneity was thought to be due to corresponding heterogeneity in factors known to vary with city, such as air-conditioning use, study population, ozone, and other photochemical oxidants.

The relationship between ozone and PM_{10} was characterized by a positive slope in summer and a negative, shallower slope in winter. The relationships between temperature and ozone were generally J-shaped. Smoothing for temporal trend did not generally alter the risk estimates.

In addition to the largely U.S.-based multicity studies (Bell et al. 2004 and Ito et al. 2005), Gryparis et al. (2004) collected data on daily ozone concentration, daily number of deaths, confounders, and potential effect modifiers for 23 European cities for at least three years since 1990. Effect estimates were obtained for each city with city-specific models and were then combined using second stage regression models. No significant effects were noted during the cold half of the year but for the warm season, an increase in the one hour ozone concentration by 10 ug/m^3 was associated with a 0.33% increase in total daily number of deaths. When considering the number of cardiovascular deaths, there was a 0.45% increase. When considering respiratory deaths, there was a 1.13% increase. The corresponding figures for 8-h ozone levels were similar.

Meta-analyses

Three meta-analyses were carried out to obtain a summary or composite estimate of ozone-associated mortality risks while explaining observed heterogeneity in risk estimates. The meta-analyses used different statistical techniques and datasets to aggregate results of multiple time-series studies of changes in ozone and mortality (Tables 4-1, 4-2, and 4-3). Ito et al. (2005) considered 43 time-series studies in the U.S. and abroad that were conducted in 1990-2003. Levy et al. (2005) considered 48 risk estimates from 28 time-series studies

TABLE 4-3 Summary of Meta-analyses

Reference	Number of Studies (or Cities)	Area[a]	Study Period	Season of Study	Exposure Metric	Lag (d)	Adjustment for Other Pollutants	Mortality	% Change in Daily Mortality per 10 ppb change in ozone (95% CI)
Levy et al. 2005	46	Inter	1973-1999	All	1-h maximum (4:3:2 1,8,24 conversion)	0-2	NA	All causes	0.41 (0.31-0.51)
	27	N. Am		All	1-h maximum	0-2	NA	All causes	0.41 (0.29-0.53)
	31	Inter		All	1-h maximum	0	NA	All causes	0.51 (0.37-0.66)
	15	Inter		All	1-h maximum	1-2	NA	All causes	0.25 (0.18-0.31)
	14	Inter		Summer	1-h maximum	0-2	NA	All causes	0.84 (0.57-1.10)
	10	Inter		Winter	1-h maximum	0-2	NA	All causes	-0.04 (-0.33 to 0.27)
	23	Inter		All	1-h maximum	0-2	NA	All causes	0.55 (0.37-0.72)
	23	Inter		All	1-h maximum	0-2			0.31 (0.20-0.45)

Ito et al. 2005	43	Inter	1990-2003	Year round	24-h average	Up to 3	NA	All causes, non-accidental	0.80 (0.6-1.0)
					Maximum 1-h average			All causes, non-accidental	0.39 (0.26-0.51)
	15	Am	1983-1999	Year round	24-h average		PM_{10}	All causes, non-accidental	0.75 (0.4-1.1)
					Maximum 1-h average			All causes, non-accidental	N/A
Bell et al. 2005	39		Various		24-h average		PM_{10} $PM_{2.5}$	All causes, non-accidental	0.87 (0.55-1.18)
								Cardiovascular	1.11 (0.68-1.53)
								Respiratory	0.47 (0.51-1.47)
					Maximum 8-h average				NA
					Maximum 1-h average				NA

[a]Inter = International; Am = America.

with study data from 1973 to 1999, excluding publications from NMMAPS. The meta-analysis conducted by Bell et al. (2005) used 144 effect estimates from 39 time-series studies published from 1990 to June 2004, also excluding studies based on NMMAPS. Each of the meta-analyses examined statistical concerns (confounding, collinearity, and possible interaction effects) related to modeling of a dependent response variable.

Ito et al. (2005). Ito et al. conducted a meta-analysis of single-city studies that investigated the short-term association between ozone and nonaccidental mortality (at all ages or over the age of 65 y). From each study, a "preselected" exposure lag (up to 3 d for consistency) was used in the meta-analyses to avoid selecting the most statistically significant lag and artificially biasing the estimate upward. Ozone-related changes in mortality risk were summarized by season. The authors conducted analyses to examine the sensitivity of the results to using different number of nonparametric smoothing terms in the generalized additive model.

The combined random-effects estimate from the 43 studies included in the meta-analysis was 1.6% (95% CI, 1.1-2.0%) excess mortality per 20-ppb increase in 24-h average ozone (roughly 0.80% excess mortality per 10-ppb increase). There was large heterogeneity in ozone-mortality risk estimates across studies that was potentially related to city-specific factors, such as mean temperature, or model specifications. Correspondingly, seasonal differences in ozone-mortality estimates were found. For the 10 studies that reported ozone risk by season, the ozone risk estimates were typically larger in summer than in winter, with random-effects estimates of 2.2% (95% CI, 0.8-3.6%) and 3.5% (95% CI, 2.1-4.9%) per 20-ppb increase in 24-h average ozone for all year and the warmer season, respectively. Although adjustment for PM_{10} had little effect on the ozone-mortality association for the whole year, the study showed potential confounding of ozone effects by PM_{10} in the warm season. That season-specific confounding by PM_{10} or temperature might explain the reported observed negative ozone-mortality associations.

Levy et al. (2005). Levy et al. applied an empiric Bayes metaregression to estimate the health effects of ozone on all-causes mortality and to assess predictors of between-study variability. Their meta-analysis included 48 city-specific relative-risk estimates from 28 studies conducted in North America and Europe and published before October 2003. In the metaregression, a hierarchic linear model with known level-1 variances was applied, and exposures were assessed with the 1-h maximum ambient ozone concentration. The meta-analysis found a significant overall 0.41% (95% CI, 0.32-0.52%) increase in daily mortality per 10-ppb increase in 1-h maximum ozone concentrations; results in North American and European cities were similar. Results from the metaregression also demonstrated seasonal heterogeneity in ozone health effects, with a significant 0.84% (95% CI, 0.57-1.09%) increase per 10-ppb increase in 1-h maximum ozone in summer and a nonsignificant -0.04% (95% CI, -0.34 to 0.28%) decrease in winter for the same increase in maximum ozone. Furthermore, results with the metaregression model suggested that between-study variability in ozone-related mortality could be partially explained by between-study differences in exposure lag times, air-conditioning prevalence, and ozone and other air-pollutant relationships, with effect estimates greatest for same-day exposures, studies conducted in cities with low air-conditioning prevalence, and a positive association between ozone and NO_2. In addition, studies using 8-h maximum ozone concentrations to estimate exposures had slightly lower estimates than studies using 1-h maximum or 24-h averages.

Bell et al. (2005). Bell et al. combined 144 effect estimates (38 from the U.S. and 106 from outside the U.S.) from 39 time-series studies to generate a pooled estimate of how ozone affects mortality. Included in the meta-analysis were peer-reviewed time-series studies that were in English, were published and indexed from 1990 to June 21, 2004, provided numerical estimates and 95% confidence intervals or t-values of the ozone-mortality relationship (total, cardiovascular, or respiratory mortality), and were not based on NMMAPS. The authors combined information across locations and estimated the pooled effect by using a two-stage Bayesian hierarchic model. Pooled estimates were generated on the basis of exposure lags, age groups, cause-specific mortality, concentration metrics, study location (U.S. or elsewhere), and potential confounding by PM. Pooled estimates from the time-series studies not based on NMMAPS were compared with pooled estimates from NMMAPS studies.

The meta-analysis showed that overall a 10-ppb increase in ozone in the few preceding days (lags of 0, 1, or 2 d or a 2-d average of lags of 0 and 1 d or lags of 1 and 2 d) was associated with a 0.87% (95% posterior interval, 0.55-1.18%) increase in total mortality. The ozone-associated risk for cardiovascular-disease mortality was larger, 1.11% (95% PI, 0.68-1.53%); and that for respiratory mortality was lower and insignificant, 0.47% (95% PI, -0.51 to 1.47%). Findings were similar for U.S. and non-U.S. cities. The statistical association was robust to the type of adjustment made for PM, weather, and seasonality. When PM_{10} or $PM_{2.5}$ was included in the model, for example, effect estimates for ozone-mediated total mortality were similar, albeit insignificant, at 0.97% (95% PI, 0.03-1.98%).

Pooled estimates were similar for the meta-analysis and NMMAPS results. Both studies showed larger effects for cardiovascular mortality (meta-analysis) and cardiovascular and respiratory mortality (for NMMAPS) than for total mortality, larger effects for exposures at lag 0 d than lag 1 or 2 d, and lack of confounding by PM. However, the estimated pooled effects from the meta-analysis were higher than those from NMMAPS, and the difference was attributed to possible publication bias (see below). The authors therefore recommend caution in the use of single-city studies, whether individual or pooled.

Summary of Multicity Time-Series Studies and Meta-analyses

Each of the multicity time-series studies and meta-analyses found strong evidence of a statistically significant association between short-term ozone exposure and mortality, and overall effect estimates were generally similar (Figure 4-1). Effect estimates, however, were higher for the meta-analyses, and this suggested the influence of publication bias on the ozone mortality effect estimates. The publication bias was attributed to the fact that meta-analyses relied solely on published studies, which are more likely to contain statistically significant findings of an ozone-mortality association. As a result, the pooled meta-analysis estimates may overstate the true ozone-mortality relationship. Moreover, inasmuch as the meta-analyses are based on individual study results rather than uniform analysis of the "raw" data from each city, it is possible that meta-analyses preferentially included the most significant results (obtained with different ozone exposure matrices and statistical models), and this, too, would result in higher effect estimates.

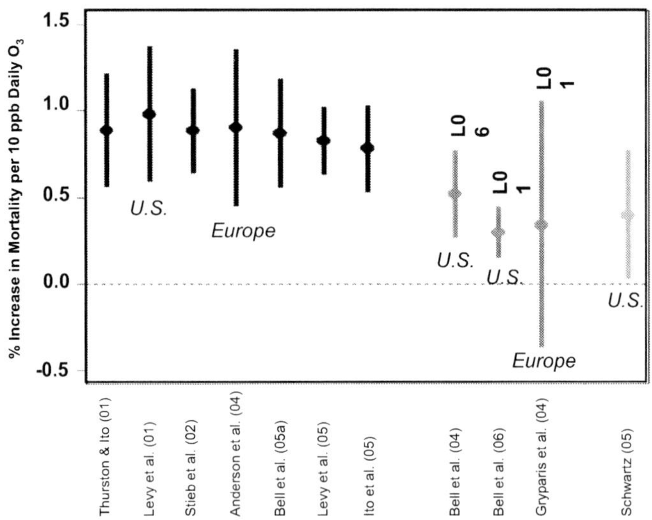

FIGURE 4-1 Percentage increase in ozone-associated mortality in meta-analyses and time-series studies. Results from all meta-analyses (*first seven bars from left*), multicity time-series (*next three bars*), and case-crossover study (*last bar on right*). "L06" refers to exposures estimated with distributed-lag models of week-long ambient ozone concentrations, "L01" to exposures estimated with average of same-day and previous-day ambient concentrations. Results presented in the papers were converted to the percent increase in mortality risk per 10 ppb increase in daily ozone levels. If both summer and yearly results were presented, yearly results were used. Several studies presented results as the 1-hour or 8-h daily maxima. To convert to the 24-hour ozone, conversion factors as noted in the studies were used when such values were provided. Otherwise, 1.73 was used for the ratio of the 1-h max to the 24-h average and 1.53 was used for the ratio of the 8-h max to the 24-h average, based on analysis of ozone levels at 78 U.S. cities. For studies that presented results based on $\mu gm/m^3$, the increment of ozone was converted to ppb based on 1.96 $\mu gm/m^3$ = 1 ppb. Source: Adapted from Bell (2007) presentation to committee. Reprinted with permission from author; copyright 2007.

Each of the investigations found heterogeneity of risk across cities or counties, particularly with regard to the time-series analyses. A recent paper by Bell and Dominici (2008) investigated whether this heterogeneity could be explained by community specific characteristics using NMMAPS. These included race, income, education, urbanization, transportation use, particulate matter and ozone concentrations, the number of ozone monitors, weather and the use of air conditioning. National relative rate did not vary greatly after adjusting for community specific variables (0.46% to 0.54% increase in mortality per 10 ppb increase in the previous week's ozone). Between-city heterogeneity in effect estimates (which is assessed in meta-analyses by using between-study variability) was attributed to several factors, consistently including seasonality and exposure lag periods. Effect estimates were generally higher in summer than in whole year and winter, irrespective of analysis or model choice, although Bell et al. (2004) found higher effects for all-year analyses compared to warm season analyses. Furthermore, ozone mortality effects were greater after same-day than after previous-day exposures in Levy et al. (2005) and Bell et al. (2004, 2005) and after exposures averaged over the week preceding death than after a single day in Bell et al. (2004). Those results suggest that ozone's mortality effects are largest on the day or days immediately preceding death but cannot be explained fully by these exposures as they accumulate over the week. Another important explanatory factor was air-conditioning prevalence: effect estimates were lower in cities that had higher air-conditioning prevalence (Levy et al. 2005; Bell and Dominici 2008).

Ozone mortality effects were generally not confounded by temperature or PM when year-round data were examined. However, the potential for confounding by PM was shown to vary by season: confounding was shown in summer but not winter by Ito et al. (2005). The seasonal confounding pattern is consistent with the seasonal pattern in ambient-ozone and ambient-PM associations in many U.S. cities. Moreover, an additional limitation of the multi-city studies is that they were not able to evaluate any co-pollutants for confounding other than PM_{10} and $PM_{2.5}$.

Although temperature was not shown to explain the ozone-mortality associations completely, the associations were found to be sensitive to adjustments for the mathematical form of temperature in models. As found by Ito et al. (2005), the use of more aggressive temperature-smoothing models resulted by itself in a 50% smaller effect estimate than other, simpler weather models. The effect of temperature adjustment was offered as an explanation for the observed doubling of the effect estimate found in the Ito et al. (2005) meta-analysis compared with Bell (2004) NMMAPS, which used four smoothing terms to adjust for temperature and dew point. Given those potentially large effects and the fact that the appropriate or best method to adjust for temperature is not known, risk estimates should consider uncertainties introduced by different model choices to account for weather. Also, care must be taken not to select the lag time with the largest risk estimate so as not to bias the estimate upwards.

Chronic Studies

Long-term effects of ozone on mortality are considered to be the result of cumulative effects on chronic pathologic conditions caused by repeated exposure to ozone, that is, not the consequence of exposure that occurred yesterday or recently but the cumulated result of conditions that develop over longer periods. These chronic conditions may lead to subclinical or clinical effects and ultimately to death. In such deaths, very recent exposure is not relevant; what matters is the long-term history of exposure that increases frailty and thus shortens life expectancy because of related chronic conditions.

The standard approach to investigate such effects is the cohort study, in which large numbers of subjects are followed for several years. Ideally, the cumulated long-term exposure of each subject to ambient ozone is estimated. The cohort study accumulates time to death. If there is no effect of ozone on mortality, subjects with high long-term or lifetime exposure to ozone will have the same life expectancy as subjects with low ozone exposure. That is, mortality rates, after all other factors that might affect mortality are taken into account, are expected to be the same among populations that have different ozone exposure histories. Cohort studies use variation in ambient air-pollution concentrations over space to create contrasts in personal exposure, whereas time-series studies use variation in time.

Chronic Ozone Exposure and Life Expectancy

Several air-pollution cohort studies have been published in North America and Europe. The mortality cohort studies have focused on ambient particulate matter or markers of local traffic to characterize exposure of the cohorts. In contrast, the assessment of the association between life expectancy and ambient ozone has been addressed to a much lesser extent. None of the cohort studies available at this time were designed to investigate chronic effects of ozone, and contrasts in estimated long-term exposure were rather limited. Results of the few cohort studies with ozone data available and that did report associations between long-term mean concentrations of ozone and death rates were not consistent.

The American Cancer Society study (ACS)—the largest cohort study of all—and the Harvard Six City study initially found a nonsignificant negative association of ozone with mortality on the basis of ozone recorded year-round (Dockery et al. 1993; Pope et al. 1995). The Health Effects Institute (HEI) reanalysis based on cohort followup from 1982 to 1989, however, reported a significant association of the third-quarter (July-September) average daily 1-h maximum ozone concentration for 1980 only and % increase in cardiopulmonary mortality of 1.026 (1.003-1.051) per 10 ppb increase in ozone (Krewski et al. 2000). The extended analysis of the ACS cohort from 1982 to 1998 (Pope et al. 2002) observed increased mortality from cardiopulmonary diseases with the 1982-1998 third-quarter average of the daily 1-h maximum ozone concentration of 1.011 % increase in mortality (0.998-1.026) per 10 ppb change in ozone. A positive but not statistically significant relative risk was observed between ozone and cardiovascular events on the basis of hospital records and deaths in the subset of 28,402 women followed for an average of 6 y in the Women's Health Initiative Observational Study (Miller et al. 2007). A negative but not statistically significant association with ozone risk was observed after adjustment for several copollutants, including $PM_{2.5}$. The 15-y followup of the Adventist Health on Smog (AHSMOG) population showed that lung cancer was significantly associated with ozone in men (Abbey et al. 1999). For other causes of death, associations were positive but did not reach statistical significance. Those few cohort studies have not demonstrated a clear positive association between long-term average ozone concentrations and cardiopulmonary mortality after controlling for $PM_{2.5}$

A persistent problem in the analyses of long-term associations between ambient ozone and health is the characterization of exposure. Factors such as correlations with copollutants, the far lower concentrations of ozone indoors than outdoors, the relevance of ventilation rates, and the influence of time-activity patterns—hamper characterization of the effects of long-term exposure to ozone. Those factors are likely sources of nonsystematic errors that are expected to lead to an underestimation of effects.

Long-Term Effects of Ozone on Lung Function

A relevant although indirect approach to evaluation of the evidence of long-term effects of pollution on life expectancy studies the effects of pollution on the development of lung function. Measures of lung function, such as forced vital capacity (FVC, the vital capacity expired during a forced spirometric test) or the forced expiratory volume (such as FEV_1, the volume expired during the first second of a spirometry test), are strong predictors of life expectancy and correlate in particular with various chronic inflammatory cardiorespiratory diseases and related deaths. Associations between long-term exposure to ambient ozone and the development of lung function throughout life can provide useful complementary evidence about the role of ozone as a risk factor for shorter life expectancy. Lung function increases during childhood and adolescence and steadily declines with age after a few years of a plateau phase in early adulthood. Reduction in growth or accelerated decline can result in clinically relevant impairment of lung function which in turn strongly correlates with premature death.

As in the case of mortality studies, PM-related investigations of lung function are more abundant and extensive than those dealing with ozone. However, some studies were either designed to or did report associations between ambient ozone and lung function in children. Recent relevant examples are the Southern California Children's Health Study (CHS) (Gaudermann et al. 2004), the University of California, Berkeley (UCB) ozone studies (Künzli et al. 1997; Tager et al. 2005), the Yale college-student study (Galizia and Kinney 1999), and a cohort study in Mexico City (Rojas-Martinez et al. 2007). Both lung function and growth correlated with ambient pollution in those investigations, but the specific contribution of ozone was less clearly established.

Interpretation of findings of effects of chronic ozone exposure on mortality is complicated by an unexplained inconsistency reported in the CHS, the largest study, with 8 y of annual followup already published. At the beginning of the cohort study, elementary-school children's lung function correlated with the long-term mean of the 1-h maximum ozone across 12 communities (Peters et al. 1999). Associations were particularly pronounced with respect to small-airway function. That would be expected in that tissue dose of inhaled ozone is known to be highest in the small airways (Hu et al. 1994; Kabel et al. 1994), but the findings were significant only in girls and in boys who spent more time outdoors. During the cohort followup, however, lung-function *growth rate* showed significant associations with a set of urban pollutants, although findings for ozone were not significant and were inconsistent among age groups and measures of function (Gauderman et al. 2000, 2002, 2004). Growth rates in small-airway function were inversely (although not significantly) associated with ozone in the youngest cohort (Gauderman et al. 2002) but not in the 8-y followup from the age of 10y to 18 y (Gauderman et al. 2004). The comparisons were based on community-level assignment of exposure to ozone, that is, spatial differences in ozone concentrations within communities and among homes were not available. As emphasized by the same research group, indoor:outdoor ratios of ozone among houses in some of the CHS communities are very heterogeneous (Avol et al. 1998), and strong negative correlations between (measured) ozone and NO_2 along busy roads have been reported (McConnell et al. 2006). In contrast to PM, indoor levels of ozone are far lower then outdoors, and the use of outdoor monitors to characterize ozone exposure has been shown to be far more challenging than in case of PM. In the main analyses of the CHS, ambient ozone has been measured only at the central monitor, so the negative correlations may be a serious problem in light of the association observed in the CHS between proximity to traffic and lung-function development (Gauderman et al. 2007). Compared to other pollutants, the cross-community contrasts in ozone concentrations was smaller and the same study found rather strong effects related to organic and inorganic acids. The formation of the latter is closely tied to ozone photochemistry. Moreover, Gauderman et al. (2004) also showed that the community (mainly Long Beach) with the lowest ozone concentrations ranked highest for elemental carbon, which correlated strongly with lung-function growth. Control of such strong negative correlations between pollutants is challenging in a 12-community design and highlights the inherent difficulties faced in all chronic-effects studies in the assessment of personal exposure to ambient ozone. As discussed elsewhere in this report, one may need an individual-level rather than community-level characterization of exposure to ozone.

Several cohort studies—including those with UCB students (Künzli et al. 1997; Tager et al. 2005) and Yale freshmen (Galizia and Kinney 1999) mentioned above—assigned individual ozone exposures for each study subject. Results of the studies suggest adverse long-term effects of ozone on lung function, which were robust in multipollutant models and among studies. The finding of the first UCB pilot study (Künzli et al. 1997), for example, was fully confirmed in a second investigation (Tager et al. 2005) with a larger sample that showed lifetime home outdoor ozone exposures significantly associated with small-airway function. The studies were cross-sectional, so their findings can be compared only with the cross-sectional (partly positive) findings of the CHS mentioned above (Peters et al. 1999).

Correspondingly, a recent followup study of a Mexican cohort of children (Rojas-Martinez et al. 2007) reported significant associations of ozone with lung growth, and adjustment for copollutants did not remove these associations. Each child lived within 2 km of a monitoring station. The stations were placed at the schools and so captured the ambient conditions during school hours and in particular exposure during times spent outdoors at school. The study was based on 36 clusters of children who shared schools and residential neighborhoods and thus had similar exposures (Rojas-Martinez et al. 2007).

Ihorst et al. (2004) repeated lung-function measurements twice a year for 3.5 y in 2,153 schoolchildren in 15 towns in Germany and Austria. They concluded that ozone may be related to seasonal changes in functional growth. No effects were detectable if the measurements were integrated over the entire 3.5 y of followup. The latter finding may be due to partial reversal of acute seasonal ozone-mediated effects on lung capacity, as described in chamber studies, but it is also of note that contrasts in the long-term concentrations of ozone were rather small among the towns.

Effects of ozone on adult lung function have been poorly investigated in epidemiologic studies. Recent studies had insufficient ranges of ambient ozone concentrations or did not use ambient ozone data at all. In a study conducted more then 20 y ago in the Los Angeles area, both lung function and its decline correlated with oxidant pollutants (Detels et al. 1987; Tashkin et al. 1994). However, although the findings were suggestive, firm conclusions could not be drawn, given that the comparison relied on only three Southern California communities; it was impossible to disentangle effects of various components of the pollution or of community-specific confounders from those expected to be due to ozone.

According to animal studies, adverse effects of long-term exposure to ozone on pulmonary function are biologically plausible. As discussed previously, chronic ozone exposure results in the formation of fewer airways, hyperplasia of bronchial epithelium, increased mucous cells, and shifts in airway smooth-muscle orientation and abundance, all of which increase airway hyperreactivity in monkeys. Lung-function decreases are both a direct result of those effects and a secondary result of the increase in insults that the lungs will experience because of increased susceptibility to infection and exacerbation of respiratory problems later in life.

It should be noted that an indirect argument for plausibility in this context leaves open the possibility for impaired lung function being only a marker of poor health rather than a definite step in the biologic pathway of chronic effects of ozone. Another possibility is that ozone could affect mortality without any effects on lung function. However, chronic subclinical systemic inflammation due to ozone may well contribute to damages in the lung or the vascular system of the lung, resulting in lower lung function, which is a typical feature of many chronic inflammatory diseases. Thus, the indirect argument for coherence linking chronic effects on lung function with long-term effects on life expectancy has some appeal despite not being stringent proof. Unfortunately, as discussed below, neither chronic effects on lung function nor on mortality are as well and consistently established in ozone studies as they are for urban air pollution marked with ambient particulate matter due the lack of studies designed for the purpose to understand chronic effects of ozone.

Summary of Chronic Effects

Some inconsistencies remain to be clarified in the epidemiologic findings regarding ozone's effects on lung function during childhood and adolescence, but the observed associations between ozone and small-airway function should not be downplayed. Ozone tissue dose is highest in the small airways, so the findings are in line with expected ozone-related pathophysiologic conditions that result in reduced lung function. The evidence of an effect of long-term exposure to ozone on lung-function growth increases the plausibility of an effect of the same exposure on mortality. The association between poor lung function and reduced life expectancy is strong and well established.

SUSCEPTIBILITY

It is well accepted that susceptibility to adverse effects of environmental exposures depend on endogenous factors (such as sex, age, and genes) and exogenous factors (such as diet, physical activity, and time spent outdoors). Such co-determinants of a person's response to ozone would lead to either attenuation or amplification of the effects. Exposure to increased concentrations of ozone has been shown to be associated with adverse health effects in some seemingly healthy individuals, as well as in individuals who are members of susceptible groups, including asthmatics (Romieu et al. 1997; Bernstein et al. 2004; Finkelstein and Johnston 2004; Maynard 2004; Wilson et al. 2004; McCunney 2005; Trasande and Thurston 2005). Studies have estimated that about 30 to 50% of the population may experience measurable adverse effects after short-term exposure to ozone at concentrations such as about 100 to 200 ppb (McDonnell 1991; Corradi et al. 2002; Arjomandi et al. 2005). Observations of different responses suggest that individual host factors, are likely to be important determinants of adverse health effects associated with ozone exposure (Kleeberger

1995; Drazen and Beier 1997; Arjomandi et al. 2005; Kleeberger 2005; McCunney 2005).

Although susceptibility varies among people, questions remain about the factors that make people more or less susceptible. Answers to those questions would be helpful in the evaluation of risks.

As summarized below, it is recognized that host factors regulate the biologic and physiologic responses to ambient ozone in rodents (Kleeberger et al. 1997; Savov et al. 2004) and humans (Weinmann et al. 1995; Balmes et al. 1996). Studies have focused mainly on acute rather then chronic effects of ozone, but modifiers of acute effects may to some extent be markers of susceptibility to chronic effects, given that some chronic pathologic conditions may be the cumulated results of repeated acute effects. The following discussion of factors in the susceptibility to acute effects of ozone may therefore be of broader relevance.

Age

Whether age modifies the effect of ozone has not been extensively investigated. Studies that have addressed the issue have suggested lower relative risks in the young. Because of the much higher baseline risk of death in the elderly, the absolute contribution of ozone may be substantially larger in them. The NMMAPS analyses of Bell et al. (2004) reported larger effects in people 65-74 y old (0.7% change per 10-ppb increase in the previous week's ozone concentration) than in younger and older people (about 0.5%). Medina-Ramon and Schwartz (in press) reported that individuals greater than 65 y old presented an additional increase in mortality as compared to younger individuals. Normal growth during adolescence and aging of the adult lung are suggested as host factors that moderate the functional changes induced by ozone exposure. A normal physiologic aspect of growth and aging is a predictable increase followed by a decline in lung function. In laboratory-based studies designed to evaluate age as an important variable in response to ozone exposure, two investigations (Drechsler-Parks et al. 1987; Reisenauer et al. 1988) demonstrated a reduced acute functional response (such as FEV_1) to ozone in subjects over 40 y old. However, the decline in the functional response with age does not necessarily imply that effects of ozone *on mortality* would follow the same pattern, inasmuch as different pathophysiologic pathways may play a role. The negative correlation between functional response to ozone exposure and age may be in part attributable to the beneficial role of the mucous lining of the conducting airways in lung defense against irritant gases (Emmons and Foster 1991).

It is important to note that early exposure to ozone may have effects over the life span of a person. Recent studies have demonstrated ozone's substantial impact on a variety of respiratory conditions in children, including the genesis of asthma in young people (McConnell et al. 2002), which may have later effects on health and pollution-mediated responses. Furthermore, the response to ozone in children is clearly different from that in adults. Children spend more time outdoors, where ozone concentrations are highest, and have higher levels of activity, which contribute to greater ventilatory rates relative to body mass than observed in adults; these factors logically are associated with greater ozone doses in children than in adults. Those conditions, coupled with the fact that children have an immature immune system and are in the process of developing the capacity to metabolize a wide variety of xenobiotic compounds, accentuate their susceptibility to environmental injury.

Studies conducted in nonhuman primate neonates (rhesus monkeys) have demonstrated that early exposure to ozone is associated with a substantial reduction in airway formation and morphogenesis that results in the formation of fewer airways, hyperplasia of bronchial epithelium, increased mucous cells, and shifts in airway smooth-muscle orientation and abundance, all of which increase airway hyperreactivity (Joad et al. 2006; Fanucchi et al. 2006). Other alterations in postnatal lung development include interrupted differentiation of airway basement membranes, modified airway epithelial nerve-fiber distribution, and reorganization of the airway vascular and immune systems (Kajekar et al. 2007; Plopper et al. 2007). Such alterations in early life can have lasting consequences, such as increased susceptibility to infection, respiratory compromise, and exacerbation of insults to the lungs later in life.

Sex and Ethnic Background

Epidemiologic studies of the acute effects of ozone on mortality have not paid much attention to sex or ethnicity as modifiers of effects. Medina-Ramon and Schwartz (in press) report that blacks are more susceptible to ozone associated mortality than non-blacks, and females are more susceptible than men. The laboratory-based studies mentioned above (Drechsler-Parks et al. 1987; Reisenauer et al. 1988) have demonstrated not only a reduced functional response (such as FEV_1) to ozone in the elderly but a larger response in female subjects than in males. In both studies, the effective ozone exposure dose (the product of concentration, duration of exposure, and minute ventilation) inhaled by the female subjects was lower than that inhaled by the males. It seems that older, even postmenopausal women are selectively more responsive to ozone exposure than age-matched men: the reasons for the difference remain unknown, but there may be interactions between hormonal factors and mortality. That has not been specifically investigated.

In laboratory-based studies (Seal et al. 1993) designed specifically to evaluate ozone-mediated responses by race and sex, the pattern of response was not significantly different among four sex-race groups (male and female, white, and black). Changes in FEV_1, airway resistance, and cough in response to ozone at various concentrations (1, 120, 180, 240, 300, and 400 ppb) were compared; a significant decrease in FEV_1 was observed after exposure to ozone at each concentration. When the FEV_1 responses were collapsed across all ozone concentrations and subjected to multiple comparison analysis, it was found that black men and black women had significantly larger decrements in FEV_1 than white men and that black men had significantly larger decrements than white women. The interrelation between acute functional response and acute risk of death was not established.

Pre-existing Diseases

For biologic reasons, it is plausible that pre-existing morbidity can modify the effects of exposure to ozone.[2] For example, a subject in the critical phase of a bout of pneumonia will have lower capacity to deal with additional exogenous stressors than will a healthy person. However, formal tests of interactions of ozone exposure with pre-existing conditions are not much published yet for ozone effects on mortality. Bell et al. 2004 observed similar effects for total mortality and death from cardiorespiratory causes. Medina-Ramon and Schwartz (in press) reported that among a range of chronic diseases considered to modify the mortality effect of ozone exposure, only atrial fibrillations showed significantly stronger associations with ozone. The additional percent increase in mortality risk per 10 ppb change in ozone was 1.66% (0.03, 3.32). Diabetes, stroke and other conditions did not modify the association. Although effects were also 1.35% larger per 10 ppb change in ozone among those with asthma listed as the second cause of death, this modification was not statistically significant (-0.31, 3.03).

The preliminary findings of mortality studies are supported by ozone studies that used outcomes other than death, such as those reporting an increase in asthma hospitalizations during days with high ozone concentrations. Subjects with respiratory diseases (such as asthma) and those prone to cardiovascular death (such as those with atherosclerotic plaque) may be considered "susceptible."

Genetic Susceptibility

Functional differences in the genes that play a role in the pathogenesis of the adverse effects of ozone are the primary candidates for gene-ozone interactions. Several studies, summarized next, confirm that concept. The issue has not been addressed with respect to acute effects of ozone on mortality.

Results from both animal and human studies support a role of genetic susceptibility in modulating ozone-induced lung inflammation and other pathologic effects and suggest important roles for genes associated with both oxidative stress and innate immunity (reviewed in Backus-Hazzard et al. 2004). In humans, genes associated with oxidative stress, redox balance, and innate immu-

---

[2]The concept of pre-existing diseases or frailty does not imply that susceptible people are necessarily very ill and destined to die soon.

nity have been examined for associations between genetic polymorphisms, oxidant exposure, and adverse respiratory outcomes. Several of those studied are related to oxidative stress and redox balance, for example, glutathione *S*-transferase M1 (GSTM1), glutathione *S*-transferase P1 (GSTP1), nicotinamide adenosine dinucleotide (phosphate) reduced: quinone oxidoreductase (NQO1), glutathione peroxidase-1 (GPX1), glutathione reductase (GR), and superoxide dismutase-2 (SOD2).

The GSTM1 null and NQO1 wild-type (NQO1wt) variants have shown interesting results. In studies in Italy, Bergamaschi et al. (2001) reported small decrements in both FEV_1 (about 3%) and peak expiratory flow (PEF) (about 9%) in 24 healthy subjects with GSTM1 null and NQO1wt genotypes after a 2-h outdoor bike ride when ambient ozone was over 0.080 ppm. Functional changes were associated with small increases in serum CC16 (interpreted as reflecting an increase in epithelial permeability). In a later laboratory-exposure study using similar ozone concentrations, Corradi et al. (2002) reported increased exhaled-breath condensate (EBC) 8-isoprostane, leukotriene B_4 (LT-B_4), and thiobarbituric acid-reacting substances (TBARs) in 22 healthy nonsmoking subjects in the same genetic groups (GSTM1 null and NQO1wt) after 2 h of intermittent exercise.

With respect to responses to ozone in the presence of both airway disease and genetic factors, Romieu and colleagues have observed asthmatic children exposed to high ambient ozone concentrations in Mexico City (Romieu et al. 2002, 2004; David et al. 2003). In about 150 asthmatic children, both decrements in forced expiratory flow rate at 25-50% lung volume ($FEF_{25\text{-}75}$) (5.2% decrease per 0.05-ppm increase in ozone concentration) (Romieu et al. 2004) and increased breathing difficulties (8% increase per 0.020-ppm increase in the ozone 1-h daily maximum averaged over 7 d) were found in GSTM1 null asthmatics; similar deficits in lung function growth were reported earlier by Gilliland et al. (2002) for asthmatic children who were homozygous for the GSTP1val 105 allele. In a case-control study of 218 case-parent triads, David et al. (2003) observed that among GSTM1 null children with high ozone exposure in Mexico City the children who were NQO1ser187ser homozygotes appeared to be protected from the development of asthma compared with NQO1pro187pro homozygotes. The finding of an interaction of ozone effects with those genes supports the notion that oxidative stress is a relevant pathway of action.

Socioeconomic Status and Other Factors

Socioeconomic status (SES) may be a marker of a variety of factors that modify the effects of ozone. Modification of effects by SES has been investigated in the case of urban PM pollution rather than ozone. There is a high correlation between SES—including diet and the intake of antioxidants, physical activity, time spent outdoors, strenuous work-related outdoor work, the use of air conditioning, stress, and exposure to environmental copollutants—and various health-relevant (both adverse and protective) effects (O'Neil et al. 2003).

The intervention study of Romieu et al. (2004) supplied antioxidants during a "treatment" period. The intervention period resulted in lower effects of ozone on lung function than in periods of supplementation with placebo. The protective effect of antioxidants was particularly strong in those genetically deficient in the ability to oppose oxidative stress (Romieu et al. 2008).

Relevance of Susceptibility to the Interpretation of Mortality Studies

The fact that occurrence or degree of adverse effects of ozone depends on other host factors has important ramifications in the interpretation of epidemiologic findings. It is relevant for risk assessors and policy-makers. The few studies summarized above lead to the preliminary conclusion that the effects of ozone on acute death rates are likely to be larger in those with pre-existing pathologic conditions and the list of plausible candidates modifying effects is rather long, although poorly investigated at this point in time (see above). Although the role of genes has not been investigated for mortality, they undoubtedly play an important role as modifiers of various pathways involved in premature mortality. Interactions among combinations of all those factors have not been investigated. Although some susceptibility factors may explain heterogeneity in effects only within populations, other marker of susceptibility may very across cities, thus partly explain heterogeneity in ozone effects observed between cities such as, for example, the use of air conditioning (Zanobetti and Schwartz 2008). Medina-Ramon and Schwartz (in press) reported that susceptibility factors had a larger effect in cities with lower ozone concentrations. For instance, the additional increase in ozone-related mortality for the elderly was 1.48% (0.81, 2.15) in a city with a mean ozone concentration of 42 ppb versus 0.45% (-0.27, 1.19) in a city with a level of 51 ppb.

One can infer that the occurrence of effects depends on subjects' susceptibility profile, which consists of a wide range and combination of single factors. "Susceptibility" is most likely not a dichotomous trait but a dynamic characteristic that follows a wide distribution in the population, from no to high susceptibility. Given the involvement of various complex mechanisms, susceptibility may be outcome-specific or pathway-specific. The level of susceptibility of any subject at any time may depend on the joint distribution of various exogenous and endogenous determinants of susceptibility (and their interactions) and may change over time. Another view of or consequence of the variation in susceptibility is that the *dose* (or the ambient concentrations) necessary to trigger an event (death in this case) and the *time* between exposure and event must follow some distribution among people.

The observable association between ozone and mortality thus reflects the weighted average of all true (unobservable) individual concentration-response functions. The latter cannot be established, but the risk of dying because of a 10-

ppb increase in ozone concentrations is likely to be substantially larger among the most susceptible population.

We emphasize that the concept of "frailty" discussed later in this chapter in the section on short-term mortality displacement can be seen as a complementary way of describing aspects of "susceptibility." Very short displacement of death (or "harvesting") may reflect the most extreme tail in the continuum of losing just a few days of life due to ozone exposure versus the loss of months or years at the other end of the distribution. Those carrying the worst combination of susceptibility factors at any point in time may be viewed as the faction in the population with the highest frailty and the shortest remaining life expectancy.

While "heavily diseased" people may belong to the "frailty pool," "high susceptibility" or "high frailty" may as well be a clinically silent condition of an apparently "healthy" subject. For example, substantial evidence supports the notion that pollutants such as urban particulate matter or diesel exhaust activate an array of partly interrelated mechanisms that are all involved in triggering myocardial infarction (MI) and arrhythmia (Brook et al. 2004; Mills et al 2007). Subjects prone to suffer an MI or arrhythmia would belong to the pool of "frail" (or susceptible) subjects, but many of those who suffer lethal or nonlethal MI or arrhythmia would not be considered "heavily diseased" until the occurrence of the first event (the MI or arrhythmia), because often the underlying pathologic condition, such as atherosclerosis, is silent. Thus, being susceptible does not mean being severely ill. It is also likely that frailty varies in any person because although acute illness (such as pneumonia) might put a person in the pool of those susceptible to death from ozone exposure, if the person survives and recovers the frailty would be decreased.

Although susceptibility factors certainly matter, the distribution of the ozone-mortality effect estimates across the categories of susceptibility is not known; that is, the quantitative details of the heterogeneity of effects are not readily available. Therefore, the overall (population-weighted average) effect in the total population is the only currently scientifically supportable entity to use in risk assessment. Its use is appropriate, even though the population-weighted mean effect may not be a valid estimate for any specific sub-population, because the 'true' effect may be much larger among the susceptible individuals but much smaller or zero among the less susceptible. To determine the change in risk associated with changes in ambient ozone concentrations for a specific location, the estimated posterior distribution of the city-specific ozone-mortality risks should be reported, including the mean and standard deviation of the distribution, and the Bayesian shrunk city-specific risks.

OTHER FACTORS THAT AFFECT INTERPRETATION OF OZONE MORTALITY EFFECTS

Errors in Exposure Estimates

The use of daily, 1-h maximum, or 8-h maximum ambient concentrations to estimate exposure in ozone-mortality studies clearly results in error. Results of exposure studies suggest that the error is large, with consistently weak (and often insignificant) associations found between 24-h ambient ozone concentrations and corresponding personal exposures and with personal exposure concentrations at or below the measurement limit of detection. As discussed in Chapter 3, those weak associations (based on the results of exposure studies relying mainly on passive ozone monitors) are attributed primarily to people's spending most of their time indoors, where ozone concentrations tend to be low and uncorrelated with outdoor concentrations. For ozone-mortality studies, which are time-series and case-crossover in design, the use of 24-h ambient concentrations to estimate exposures will result in reduced power to examine associations between ozone exposure and mortality (Carroll et al. 1995; Zeger et al. 2000) and can lead to bias in estimates of ozone mortality effects (Zeger et al. 2000).

The magnitude of exposure errors probably depends on several factors, including season, home ventilation characteristics, and exposure averaging time. Ozone's mortality effects, for example, were shown to be greater in the warm season than in winter or the entire year (Ito et al. 2005; Levy et al. 2005). Greater ozone effects in the warm months are consistent with people's spending more time outdoors and tending to live in homes that have better ventilation in the warm months. As evidenced by results of exposure studies, both those factors should lead to stronger 24-h ambient-personal associations, and thus lower exposure error, in the warm seasons, through increased contributions of ambient ozone to personal exposures and through greater infiltration of ozone from outdoor to indoor environments, respectively. Those factors may also result in effective doses that vary by season, for example, because of differential physical-activity patterns and breathing rates, which would result in seasonal differences in exposure error. The seasonal differences may be mitigated by behavior patterns that are modified by air-quality alerts, which lead people to spend more time indoors on high-ozone days. The effect of behavior during ozone air-quality alerts warrants further examination, as discussed in Chapter 3.

The importance of season and ventilation has been further shown by Levy et al. (2005), who provided preliminary evidence of decreasing city-specific ozone mortality effects with increasing prevalence of central air conditioning, presumably because of weaker personal-ambient ozone associations and thus greater exposure error in homes with air-conditioning use. Further support is provided by Ito et al. (2005), who found mean city-specific temperature to be negatively associated with corresponding city-specific ozone-mortality risk estimates, possibly because of the influence of air-conditioning prevalence in the cities. Those findings suggest that error due to the use of 24-h ambient concentrations as the exposure metric is greater in winter than in summer, and this offers a potential explanation of the lower ozone mortality effects in cold seasons than in warm seasons.

Even in the warm months in non-air-conditioned homes, when exposure errors are lower, it probably remains a concern for health studies and their interpretation. As discussed in Chapter 3, when 24-h ambient concentrations are significantly related to 24-h personal ozone exposures, the slope of their association remains small and substantially lower than those found for $PM_{2.5}$ and sulfate (Sarnat et al. 2000; Sarnat et al. 2006). The low slopes may raise questions about biologic plausibility. Taken at face value, they suggest that some minute change in the true (unknown) 24-h mean exposure to ozone of only 1 ppb or a few parts per billion could cause death. That appears hard to accept, but other interpretations of the studies may lead to less paradoxical conclusions.

Given the observed consistency in ozone-mortality findings and the low 24-h personal-ambient ozone associations, it is possible that 24-h ambient ozone concentrations are representative not to 24-h personal ozone exposures, but to some correlated toxic agent or agents. These correlated toxic agent(s) could include photochemical pollutant mixtures, ozone byproducts, or ozone exposure of shorter or longer duration. For exposures of shorter duration, for example it is well established that 24-h ambient concentrations are strongly correlated with 1-h and 8-h maximum ambient concentrations.

Although it has not been much studied, it is also possible that maximum ambient ozone concentrations averaged over 1-h or 8-h are better proxies (with respect to both strength and magnitude of association) of corresponding 1-h and 8-h ozone exposures than 24-h ambient concentrations and exposures. In a scripted-exposure study using a trained technician, for example, 1-h ambient concentrations were strong proxies of corresponding personal exposures when the technician spent the hour outdoors. During the time outdoors, 1-h ambient ozone concentrations explained most of the variability in personal exposures and were about equal to personal exposures (Chang et al. 2000).

The generalizability of those study findings to at-risk populations, such as the elderly, after their typical activities is not known and warrants future study. People are outdoors primarily when maximum 1-h and 8-h ozone concentrations occur in the afternoon, so it is possible that ozone-mortality studies are reflecting maximum ozone exposures averaged over 1 h or 8 h rather than the daily ambient ozone concentration. Given the high correlation between these ambient short-term concentrations and the 24-h mean associations, use of the 24-h concentrations in time-series studies may serve as an equally good proxy for peak exposures. If so, the differences in the true (unknown) day-to-day peak exposure may indeed be far more than 1 ppb or a few parts per billion and reach some 20-40 ppb. Such an interpretation would resolve the apparent paradox of a few parts per billion causing death.

Confounding

Co-linearity among ambient pollutants raises concerns about possible confounding of ozone mortality effects by correlated co-pollutants, such as $PM_{2.5}$ (and its components), NO_2, and SO_2. This concern centers upon the possibility that ambient effects associated with ozone concentrations may represent not only consequences of ozone but also of correlated pollutants not included in the health effects model. This possibility is critical to address, as our understanding of confounding is central to our ability to ascertain the effects of ozone on mortality and any benefits that will result from ozone control. Of the possible confounders, ambient $PM_{2.5}$ its components, and weather have raised the most concerns about confounding. This concern results from several factors, including the strong correlation of ambient ozone with temperature and ambient $PM_{2.5}$ as well as between mortality and both temperature and ambient $PM_{2.5}$. For $PM_{2.5}$, concerns over confounding also arise from the relatively weak association between personal and ambient ozone, especially in relation to the much stronger association between personal and ambient $PM_{2.5}$. As noted in Chapter 3, relations between ambient ozone, temperature and ambient $PM_{2.5}$ vary substantially by season and geographical location, suggesting that the potential impact of confounding must also be considered by season and location. Also, as noted in Chapter 3, the association between ozone and $PM_{2.5}$ components is less well understood, in particular as to how this association varies by location, season, and component and how it differs for ambient concentrations and actual personal exposures.

Even with these seasonal and geographical considerations, controlling for confounding in ozone mortality studies is a considerable challenge. To date, ozone mortality studies generally have addressed confounding by weather (generally temperature and/or dew point) using a variety of approaches, including those based on indicator variables, linear terms, V-shaped linear terms, and non-linear smooth terms. The ability of these approaches to control for confounding and their corresponding impact on ozone mortality risk estimates have been shown to differ. In a reanalysis of data from seven cities, for example, Ito et al. (2005) examined the impact of control for weather using four approaches (quintile indicator variables, V-shaped linear terms, and two combinations of non-linear smooth terms) and showed that overall ozone mortality risk estimates differed by as much as 100% when different approaches were used to control for weather. Models using indicator variables had the highest risk estimates, while models with four non-linear terms had the lowest estimates. Further, the model with the four non-linear terms had the best statistical fit but also the greatest concurvity with ozone, suggesting that ozone and temperature effects from these models are the most difficult to separate. Consistent with these findings, Levy et al. (2005) (using a meta-analytic approach) found that studies with non-linear temperature terms had higher ozone mortality effect estimates than studies using linear terms, although results were limited by the fact that only a small fraction of the examined studies used linear terms. Despite non-linear terms having the lowest effect estimates, Levy et al. (2005) and Ito et al. (2005) are consistent in that the ozone risk was influenced by the manner in which temperature was modeled. Implications of these findings to other ozone mortality studies are not known. At a minimum, the observed large impact of model choices on ozone mortality risk estimates suggests that control for weather is an important source of uncertainty in ozone mortality risk estimates.

To address uncertainties based on model choices, Schwartz (2004) examined ozone mortality effects in 14 cities using a case-crossover approach, which controlled for temperature mainly by matching control days with the temperature on the day of death. Using this approach, Schwartz showed a significant overall effect of ambient ozone on mortality that was greater in warm as compared to cold seasons, with these effects relatively unchanged when control days were matched on temperature. Similar effects when control days were matched or unmatched on temperature suggest that observed ozone mortality effects are not due to confounding by weather. However, these findings need to be replicated in additional studies. Further research examining the potential for confounding by weather and the impact of methods used to control for this confounding are clearly needed.

For $PM_{2.5}$, ozone mortality studies have addressed concerns about confounding focusing on $PM_{2.5}$ as mainly as a whole (and not its components). The potential for confounding has been examined using multiple pollutant models, showing for example that yearly ozone mortality estimates are robust to adjustment for ambient particles and suggesting that confounding of ozone mortality effects by ambient particles is not substantial (Bell et al. 2007). Interpretation of the findings from these multi-pollutant models, however, is complex. For example, the use of two-stage regressions (as was used in the ozone meta-analyses) to examine confounding by correlated pollutants have raised statistical concerns about the potential for misleading results (Marcus and Kegler 2001). Further, measurement error can differ substantially by pollutant, raising further issues regarding interpretability of findings. Franklin and Schwartz (2008) found a 0.89% (95% CI: 0.45%, 1.33%) increase in non-accidental mortality with a 10 ppb increase in same-day 24-h summertime ozone across the 18 communities. After adjustment for PM2.5 mass or nitrate this estimate decreased slightly but when adjusted for particle sulfate, the estimate was substantially reduced to 0.58% (95% CI: -0.33%, 1.49%). The authors concluded that the association between ozone and mortality is confounded by particle sulfate, suggesting that some PM components could be responsible for part of the observed ozone mortality effect.

Interpretation of multi-pollutant results is further complicated by the strong correlations among the ambient pollutants, which make their associations with health effects difficult to separate. For multi-pollutant analyses based on yearly rather than seasonal data, this separation is particularly difficult, given the seasonal variability in ambient pollution relationships (see Chapter 3). Further, correlations with ambient ozone in the eastern U.S. generally are strongly positive in the summer but weak or negative in the winter. Evidence from Ito et al. (2005), which examined associations for seven cities located primarily in the eastern U.S., suggests that this seasonal variability in ambient pollutant correlations is important, as ozone mortality estimates for the warm and cold seasons were lower and were more similar by season when measured ambient fine particles were included in the season-specific models. Since the relation between ambient ozone and ambient $PM_{2.5}$ differ substantially by season and location, the observed similar warm and cold season associations suggest observed ozone mortality effects in the eastern U.S. were not due to confounding by ambient particles. The observed seasonal differences demonstrates the importance and need for season-specific analyses to examine potential confounding of ozone mortality effects by $PM_{2.5}$ and other pollutants. These types of analyses should be repeated for other regions of the United States.

The implications of findings based on multipollutant ambient concentrations for actual exposures and their impact on ozone mortality estimates is not clear. When ambient ozone and $PM_{2.5}$ concentrations are included in the same model, they are assumed to be proxies of their own exposures. However, results from exposure studies suggest that this assumption may not be valid in some geographic areas, as 24-h ambient ozone concentrations in the eastern U.S. have been shown to be better measures of exposures to $PM_{2.5}$ than to ozone (Sarnat et al. 2001), while 24-h ambient $PM_{2.5}$ concentrations have been related to personal exposures to ozone in addition to $PM_{2.5}$. These findings from the U.S. East Coast may suggest that ambient ozone and ambient $PM_{2.5}$ are serving as proxies for the same pollutant or pollutant mixtures, although results are somewhat limited by the fact that 24-h personal ozone exposures measured in these studies are often near or below the method limit of detection. Despite this, 24-h personal ozone exposures have been shown to be poorly correlated with corresponding personal $PM_{2.5}$ exposures, suggesting that if ozone mortality studies were to include more direct measures of ozone and $PM_{2.5}$ exposures, issues related to confounding of ozone mortality effects by $PM_{2.5}$ may be better resolved—at least for the assessment of daily exposure periods. For shorter exposure averaging periods, confounding of ozone mortality effects by $PM_{2.5}$ may still be possible, as 1-h personal ozone and $PM_{2.5}$ exposures were strongly correlated in one study when the technician was outdoors and away from roads (Chang et al. 2000). Whether exposure study results from the East Coast and for $PM_{2.5}$ can be generalized to other locations and to specific $PM_{2.5}$ components, respectively, is of yet unknown.

Air-Pollutant Mixtures

Results of toxicologic studies provide conflicting evidence on whether the mortality risks posed by ozone differ in the presence of other air-pollutant exposures. Follinsbee and coauthors (1981) have shown in young healthy subjects that exposure to a gas-phase mixture of ozone and NO_2 induced about the same degree of pulmonary-function decrease and breathing-pattern changes as were observed previously in a separate dataset on subjects exposed to ozone alone (Follinsbee et al. 1977).

In contrast, a mixture of ozone and NO_2 was found to elicit greater pulmonary-function decrements in both older (about 50-75 y old) and younger subjects than ozone alone (Drechsler-Parks et al. 1987, 1989). Older subjects were less responsive than younger subjects, but both cohorts exhibited a fairly broad range of pulmonary-function changes. Exposure to the mixture, but not exposure to the individual pollutants, led to decreases in the older subjects in stroke volume, heart rate, and cardiac output during exercise periods of the exposure. That observation suggests not only that exercise performance may be limited by exposure to the pollutant gases as a mixture but that it may put people with preexisting cardiovascular disease at increased risk of exacerbation of their disease. Cardiac-function was not studied in the younger cohort. Healthy adults (56-85 y old) showed significantly smaller exercise-induced cardiac output with exposure to the combination of nitrogen dioxide (NO_2) + ozone than with exposure to ozone or filtered air alone (Drechsler-Parks 1995). Also, as discussed earlier in this chapter, Levy et al. (2005) found the estimates of ozone-mortality effects estimates were higher where there was a positive correlation between ozone and NO_2.

Controlled exposure to mixtures has also been evaluated in susceptible populations. For example, exposure of mild atopic asthmatics for 3 h to ozone at 200 ppb, NO_2 at 400 ppb, or a combination of the two pollutants significantly increased sensitivity to inhaled allergen compared with exposure to air, without additive effects. Restudy of the same subjects for 6 h at half the concentrations did not have significant effects. The results suggest that the pollutant-induced changes in sensitivity to allergen in mild allergic asthmatics may depend on a threshold concentration of gas-phase pollutants rather than on the total amount of pollutants inhaled (Jenkins et al. 1999).

Gas-phase copollutants can exert their own direct effects on the respiratory tract; however, they may also influence the deposition fraction of solid-phase air pollutants—that is, fine and coarse particles—and localize and intensify injury to epithelial tissues. For example, ozone-induced airway obstruction has been shown to develop differentially even in healthy subjects (in a current database on 140 subjects exposed to ozone in controlled chamber studies), and as many as one-third of the subjects have a significant degree of airflow obstruction during and immediately after short periods of exposure to a concentration similar to an ambient concentration of ozone. Ozone-induced airflow obstruction has been shown to increase bronchial deposition of respirable particles ($PM_{2.5}$) in the lower respiratory tract (Foster et al. 1993). The pulmonary inflammation and increased epithelial permeability that follow ozone exposure may also lead to increased transepithelial transport of deposited particles, which may increase the risk of PM-related systemic effects.

Just as ozone can have direct effects on airway epithelial tissues, $PM_{2.5}$ exposure is known to induce systemic inflammation and proinflammatory cytokine production that may result from free-radical activity of components in PM. Controlled chamber studies have also been evaluated with 2-h exposures to concentrated ambient fine particles (at about 150 $\mu g/m^3$) in combination with ozone at 120 ppb. In healthy subjects, those exposures induced an immediate alteration in vascular function as indexed by brachial arterial vasoconstriction (Brook et al. 2002). That suggestive study has shown that acute exposure to fine-particle and ozone mixtures at concentrations often found in the urban environment can lead to acute peripheral arterial constriction.

Thresholds

To summarize the association between daily variations in ambient ozone concentration and daily fluctuations in deaths, a linear model is used in which it is assumed that the change in mortality risk is constant across levels of pollution. It is unlikely that the association between exposure and response at the individual level follows that simple mathematical formulation. People have their own susceptibility, which is characterized by a unique exposure-response association. The association may be further characterized by a unique "threshold" value, an ozone exposure concentration below which there is no risk of death. Each person's threshold value will also vary over time, depending on the frailty of the person at any given moment, and thresholds may depend on the averaging period used to assess exposure.

The time-series design, however, relates individual exposure not to individual risk but to the average of personal exposures among the at-risk population on any given day. Thus, the appropriate concentration-response function for the time-series studies is based on an aggregation of individual exposure-response curves. Aggregation of a large number of complex functions can yield smoother and more nearly linear curves at the population level, so we would expect the shape of the concentration-response function based on time-series studies to have a form that is relatively simple.

The shape of the concentration-response function for ozone and mortality has been examined in detail (Bell et al. 2006). The association between a 2-d moving average of daily average ozone concentrations in 98 U.S. communities based on time-series data from 1987-2000 was examined. Bell et al. restricted their analysis to all nonaccidental causes of death and the entire year of observation. Three methods were considered to examine the shape of the concentration-response function: restricting ozone concentrations below specified values, postulating a threshold function with no risk below the threshold and a linear association above it, and a natural spline formulation of association. A positive association between ambient daily average ozone concentrations and mortality was observed down to 15 ppb for the restriction and spline approaches, and the threshold model could not show any substantial improvement in statistical fit down to concentrations of 5 ppb, the lowest value examined. We note that positive and statistically significant ($p < 0.05$) associations were observed down to 25 ppb using the restriction approach. The risks based on 24-h mean ozone concentrations below either 20ppb or 15ppb were positive, but did not reach formal statistical significance. Threshold values based on 8-h running daily maxima would be somewhat higher.

Those results suggest a near-linear association between ambient concentration of 24-h mean ozone and daily mortality count in the United States. Misclassifying exposure (by using an average of ambient fixed-site monitoring data to estimate population average personal exposure) makes it more difficult to distinguish between linear and threshold models (Cakmak et al. 1999). Estimates of the concentration-response curve based on epidemiologic studies with imprecisely measured exposure should therefore be viewed with caution. A sensitivity analysis of the shape of the curve may be required to capture the uncertainties in this procedure more fully. Moreover, approaches based on 24-h averaging may cloud thresholds related to actual exposures which may be better represented with shorter averaging times, such as by using an 8-h maximum or daily 1-h maximum.

The analysis of Bell et al. (2006) examined data from the entire year. However, it is not clear whether there is an association between ozone and mortality in the cooler months (Levy et al. 2005). The warmer months should therefore be examined separately. There is likely to be less exposure error during that time because people tend to spend more time outdoors or with their windows open. However, the magnitude and nature of the error could depend on location. Furthermore, confounding by other pollutants, particularly $PM_{2.5}$ in the warmer season (Levy et al. 2005), may alter the shape of the concentration-response curve.

Future approaches to further examining concentration-response relationships should consider using alternative algorithms for computing nationally averaged effects, using regional nonlinear or threshold estimates, and repeating all analyses using 8-h and 1-h maxima, instead of 24-h ozone data.

On the basis of its review of the evidence, the committee concluded that the association between ozone concentration and mortality is generally linear throughout most of the concentration range although a number of uncertainties make it difficult to know whether there is a threshold for the association at the lower end of the range. If there is a threshold, it is probably at a concentration below the current ambient air quality standards.

Short-Term Mortality Displacement

Investigations of acute effects of pollution assess the association between the daily number of events and ambient air quality during the period shortly before death. By the design of the studies, the time of life lost or the "prematurity" of the deaths is not directly observable. In the absence of knowledge about the amount of time lost because of these acute effects, one may hypothesize that air pollutants were able to trigger death only among a pool of very frail people, namely, those already in very bad health. Assuming that the remaining life expectancy among those frail subjects was short even in the absence of pollution, the effect of air pollution would consist of only a minor shift of the time of death, that is, a short "advancement" of death. The concept that air pollution affects mortality only in frail people already near death is referred to as short-term mortality displacement.[3] A period with increased mortality would be counterbalanced by a period with lower then expected mortality because deaths among the pool of susceptible people have already occurred (the "premature deaths" have been displaced by a few days). It has not been defined how "short" the advancement of death must be to qualify as short-term displacement, but the committee assumes that this concept implies losing a few days rather then weeks or months.

The issue of short-term mortality displacement is important in the judgment of the public-health relevance and quantification of the public-health impact of the acute effects of air pollution if one is willing to weight the deaths of people destined to die within a very short time differently from the deaths of people who would have a longer life expectancy if there had been no exposure to ozone.

The significance of effects will be determined by the size of the frail pool and the extent of displacement among the pool (Fung 2007). Air pollutants may affect the pool in various ways. The size of the pool, the rates at which people join or leave the pool, and the causes of their joining or leaving all affect the dynamics in mortality patterns. Ozone could increase the death rate outside the risk pool, increase the recruitment rate into the pool, or delay the recovery rate outside the pool. The net acute impact would depend on the relative size of each effect, so the temporal pattern of increased deaths after exposure might be complex. If ozone affects only the death rate outside the risk pool, one would expect to see fewer people die after an ozone episode because the risk pool is smaller. If instead ozone increases recruitment or delays recovery, the risk pool may enlarge, and the number of deaths in the period after the episode would be larger rather than smaller. Rabl (2006) has emphasized that the main effects in the mortality-displacement hypothesis are in essence unobservable because observed death rates are the net result of all the changes that affect the dynamics of the pool of frail subjects. He has shown that the increase in rates due to premature displacement and the decrease in deaths after the depletion of the susceptible pool cannot be observed or disentangled.

Short-term mortality displacement has been investigated (with ambient PM as the marker of daily pollution) in several ways with a variety of statistical methods, including assessment of the displacement with generalized additive models (Zanobetti 2000), decomposition of the time series into time components (Zeger et al. 1999; Schwartz 2000), Gaussian-state space models (Murray and Nelson 2000), frequency-domain log-linear regressions (Kelsall et al. 1999), and time-scaled Fourier analyses (Dominici et al. 2003; Fung et al. 2005). Questions remain about the methods and assessment of mortality displacement, and the time lost because of acute effects of ambient PM cannot be quantified, but the

[3]The term *harvesting* is sometimes used instead of *short-term mortality displacement* to refer to the concept that air pollution leads to the death of people who are highly susceptible and near death (and die a few days earlier than they would have without air-pollution exposure) rather than death of people who are not otherwise near death.

main conclusion from all the work is that the observed acute effects cannot be explained only by short-term displacement of death.

Apart from those findings, one should emphasize that the idea that all acute effects of ozone can be explained by short-term mortality displacement involves a biologically implausible dichotomization of the distribution in the time between some adverse acute exposure and death. Instead, the time between exposure and death and the time lost are expected to follow some distributions that depend on the pattern of susceptibility factors or the frailty of the population. For example, survival time after an MI follows a wider distribution. Some MIs are immediately lethal, but survivors experience a wide distribution of increased risk of death as compared to those who have not suffered an MI, and risks are particularly increased at the onset of an MI but also in the early phase of recovery. Thus, if air pollution triggers MIs, the natural distribution of time to death after an MI should most likely be mirrored in MI-related deaths due to air pollution in that the severity of an MI and its survival pattern depend less on the trigger than on the underlying pathology, such as location or degree of atherosclerosis.

A similar case that demonstrates the inherent limitation of the hypothesis of short-term mortality displacement can be made for the frail pool of patients with chronic obstructive pulmonary disease (COPD). Air pollution may trigger acute exacerbations due to, for example, the interference of particles with defense mechanisms in the lung, but whether and when decompensation and death occur depend on various host and exogenous factors, including type and time of access to health care, that affect the distribution of probability of and time to death. During an exacerbation of COPD due to, for example, a pneumonia, the patient may be particularly frail during the critical phase of the pneumonia (typically at the end of the first week) and have a transient inability to deal with the additional stress of increased ozone exposure. That may lead to death; but in the absence of the additional hazard posed by ozone, the patient might have survived and lived for as long as an "average" COPD patient of the same age. This example underscores that frailty is not necessarily a stable condition but may change substantially over short periods in a given subject. Some of the acute effects of ozone (see earlier discussion in this chapter), in fact, contribute to moving subjects to a higher level of frailty.

In contrast, the terminal phase of lung cancer may serve as an example in favor of air pollution's being in the role of short-term mortality displacement. As shown by Goldberg et al. (2001) in patients with lung cancer, death rates correlate with ambient concentrations of urban air pollution. However, lung cancer continues to be by and large an incurable disease with short survival time after diagnosis. Thus, the role of acute exposure to air pollution in the terminal phase of this disease may be considered less relevant despite some additional displacement of lung-cancer death due to air pollution.

The above examples and discussion explain well the observation made in the literature on short-term mortality displacement by PM that risk estimates may increase if the statistical models take extended lag periods into account.

The observations after the classical London smog further support the concept of distributions of events with long tails of subacute effects on mortality rates. The latter remained above expected values for several months after the extreme smog episode without observable evidence of short-term mortality displacement.

One published study (Zanobetti and Schwartz 2008) addresses this important issue related to the acute effects of ozone. It investigated the effects of short-term ozone exposure on mortality in summer by using distributed-lag models. With that seasonal restriction, it was not possible to assess the longer tail of the effects beyond 21 d. As in the case of PM, the study does not support the notion that only short-term mortality displacement occurs. However, in contrast with the evidence from PM studies, these preliminary results suggest that death rates normalize within about 1 wk, that is, faster than in the case of PM. Accordingly, the effect estimates (or odds ratios) of distributed-lag models were not much larger than those of the acute 1- to 2-d model. As mentioned, the details of the interrelated effects of the size of the frail pool and changes in size due to persons becoming frail or recovering and leaving it cannot be disentangled in these approaches. The apparent difference between results of the PM distributed-lag models and those in the ozone studies is difficult to interpret. The 20-d distributed-lag effect of ozone on mortality was at most some 60% larger than the immediate lag-0 effect—a substantially smaller difference than that reported for PM. There is a need for further investigations of short-term mortality displacement and the distribution of lagged subacute effects of ozone, but one can conclude, on the basis of the preliminary analyses and on conceptual grounds, that acute effects of ozone based on the published meta-analytic slopes underestimate to some degree the total acute and subacute effects. The underestimation may be smaller than in case of PM.

It is also of note that the preliminary findings regarding the probable pattern of acute and subacute effects of ozone on mortality do not provide further information about the occurrence or size of chronic effects.

On the basis of the above background from the PM literature, the biologic concepts of acute effects, and the preliminary results from Zanobetti and Schwartz (2008), one can conclude the following:

- Short-term displacement of death is a likely explanation for the subgroup in the upper tail of the susceptibility of the frailty distribution. The size of the fraction explained by short-term displacement is not known.
- Short-term mortality displacement is not a plausible explanation of all effects. In fact, the short-term-displacement-only hypothesis conflicts with the notion that underlying mechanism, frailty, severity of outcomes, and time between exposure and event all follow some *distributions* in populations. Although the amount of life time lost is not observed in acute-effects studies, the upper tail of the distribution of time lost may be months or years.

The distributed-lag effect estimates from time-series studies may be used to estimate the population average lifetime lost because of acute and subacute effects of ozone if one assumes a common hazard or probability of instantaneous death in all members of the population of any given age. More likely, though, hazard rates follow a wide distribution, thus the assumption reflects a simplification. That "population mean" approach, however, does not reflect the distribution of the amount of time lost because of the deaths that occur as a consequence of increased ozone concentrations or the distribution of pre-existing frailty among those affected. These issues highlight the difficulties in translating the time-series findings into an estimate of life time lost. Sensitivity analysis should be conducted on the number of degrees of freedom used to analyze the time trend in models with distributed lags because the risk estimates for the longer lagging times can be influenced by how temporal trends in mortality effects are modeled.

Uncertainty and Variability in Ozone-Mortality Analysis[4]

One of the major barriers to the broad acceptance of recent EPA health-benefits analysis is the large amount of inherent uncertainty. How the agency deals with that uncertainty is therefore critical for acceptance.

Assessment of uncertainty is not the same as assessment of variability. Uncertainty is a consequence of imperfection in knowledge or data and can (in theory) always be reduced by getting better data. Variability is an inherent property of an exposed population and cannot be changed by getting better data. A fair and balanced characterization of uncertainty in risk estimation is important because most risk estimation is not highly precise and many people are tempted to over interpret the resulting values. EPA (1997c) recognizes that well-performed uncertainty analysis helps decision-makers and the public to place risk estimates in the proper perspective and facilitates informed decision-making. The need for uncertainty analysis is also recognized in other countries as evidenced by the European Commission report (Bickel and Friedrich 2005) that emphasizes the importance of proper communication of uncertainties to ensure that users understand the limitations of analyses and their results.

Other reports discuss the issue of uncertainty in risk assessment and benefits estimation (NAE 1972; NRC 1975, 1982, 1983, 1994, 1996, 2002, 2007a; PCCRARM 1997). All those reports found that proper and adequate characterization of uncertainty is essential. Almost all expressed concern that most risk assessments and health-benefits analyses tend to underestimate uncertainties and leave decision-makers with unwarranted confidence in the risk estimates provided. To address that concern, the reports recommended the use of formal approaches to characterize uncertainty, such as Bayesian analysis or Monte Carlo analysis (Gilks et al. 1996). Less consistent were opinions about characterization

---

[4]This section concerns the uncertainty and variability in the analysis of mortality associated with ozone exposure. The potential uncertainty in the monetary valuation of mortality risk associated with ozone is considered in Chapter 5.

of model uncertainty: some reports discussed and recommended the use of expert judgment, and others recommended that such scientific uncertainty be thoroughly described but not quantified. Efforts to characterize model uncertainty, however, are critical, as evidenced by the most recent report (NRC 2007b), which states that model uncertainty in particular tends to be understated or ignored. That report states that modelers often assume that their models are correct and base estimates of the models' parameter values on single studies; because a model may be incorrect or incomplete, the uncertainties that they produce may be significant.

Sources of Uncertainty

There are many sources of uncertainty in mortality-risk assessment related to ozone exposure, including random sampling error in a random sample of data, measurement error (systematic error or random error), data nonrepresentativeness, surrogate data, lack of relevant data, problem and scenario specification, and model uncertainty. Of those, only sampling error is captured in ordinary statistical measurers of uncertainty (p-values and confidence bounds), so total uncertainty is necessarily greater and perhaps much greater than can be directly measured.

Several methods exist to reduce uncertainty, including iteration of model-building and input distributions, in combination with sensitivity analysis, which can help to focus resources on the most important model inputs and components. The uncertainty sources that probably have the greatest influence in ozone-mortality analysis are the epidemiologic models. Much of this uncertainty results from unavoidable and expected variability, from estimation of concentration-response functions with epidemiologic studies, from possible lagged effects, and from baseline statistical variation. Baseline statistical variation, which tends to be inherent in data, could be reduced in some cases with control measures and in others with better models. To deal with model uncertainty, it is possible to compare alternative models but not combine them, weight predictions of alternative models (for example, with probability trees), or use metamodels that degenerate into alternative models. Uncertainty is also associated with ozone concentrations and with the reliability of ambient ozone-monitoring data to indicate ozone exposure.

As stated above, the formal approaches to uncertainty analysis in mortality models include Bayesian analysis and Monte Carlo analysis. The use of expert judgment, which could be considered an empiric Bayesian approach, provides another framework for uncertainty analysis. EPA used expert elicitation in the 2006 final regulatory impact analysis for the PM NAAQS (EPA 2006). However, the best way to use expert judgment remains to be determined. A key step in moving forward would be to agree on conditions under which expert judgment is an acceptable input, even for rule-making situations. The World Health Organization (WHO 2000) has developed guidelines to identify a set of processes and general approaches to assess available epidemiologic information in a clear, consistent, and explicit manner. In particular, it recommended the use of expert assessment according to an explicit protocol, defined in advance. The essential components of the protocol are: specification of the question to be addressed, justification of the expertise represented, and specification of the methods to be used. Before EPA makes a decision about the use of expert judgment in the context of the impact of ozone on mortality and for benefit analysis, further evaluation is needed to assess the relevance and usefulness of the recent expert elicitation results for PM. In particular, in view of the time and cost associated with an expert-judgment procedure, the agency needs to understand whether, in the case of PM, the expert elicitation in the presence of other sensitivity analysis added substantial value or could have been replaced with a published formal or "qualitative" meta-analysis.

EPA could also consider other options for incorporating expert judgment into its probabilistic uncertainty analysis. The agency possesses considerable internal expertise, which should be used as fully as possible. If it continues to use expert elicitation for uncertainty analysis, it should consult outside experts as needed, individually or in panels. The experts and the rationales and empirical bases of their judgments should be made known.

It is important to distinguish between uncertainty due to projecting the future and uncertainty inherent in estimating mortality on the basis of information on the magnitude of ozone exposure, which varies across space and time and among individuals.

Thus, a good characterization of temporal and spatial correlation is also needed. Sensitivity analysis and model diagnostics would help to determine the appropriateness of the spatial and temporal characterization of the dependence structure (and other assumptions) incorporated in the models under consideration.

The ability to quantify and propagate uncertainty throughout the analysis is still in development. However, uncertainty analysis has developed further and faster than our ability to use it in decision-making. A National Research Council report (2007a) recommends that more attention be paid to questions of how to use uncertainty analysis to set action levels and make regulatory decisions. Effective communication of uncertainty requires a high level of interaction with the relevant decision-makers to ensure that they have the necessary information about the nature and sources of uncertainty and its consequences.

Uncertainty in Air-Quality Numeric Models

Air-quality numeric models, such as the Community Multiscale Air Quality (CMAQ) model, are potentially valuable tools that can extend the ozone-mortality analysis to places and times on which the desired data are not available. Air-quality models, which are based on the dynamics and mechanics of atmospheric processes, typically provide information on larger regions than data from observational networks. Errors and biases in these deterministic models are still inevitable because physical processes are simplified or neglected and because of the mathematical approximations used in assigning parameter values and inaccurate inputs. As a result of their own uncertainties, the models will add more uncertainty to the ozone-mortality risk-assessment estimates if they are used to characterize ozone exposures in future studies.

Evaluation of these models and their uncertainties can help to quantify and characterize the magnitude of errors in the models. Although a full Bayesian analysis that incorporates all sources of information may be desirable in principle, it will be necessary in practice to make strategic choices about which sources of uncertainty justify such treatment and which sources are handled better through less formal means, such as consideration of how model outputs might change as some of the inputs vary through ranges of plausible values.

A National Research Council report (NRC 2007b) describes in more detail different sources of uncertainty in air-quality numeric models and addresses the question of how to judge whether these models and their results are adequate for supporting regulatory decision-making. Effective communication of the different sources of uncertainty in the models requires a high level of interaction with the relevant decision-makers to ensure that they have the necessary information about the nature and sources of uncertainty and its consequences. Thus, if such models are used to assist in mortality risk assessment, uncertainty analysis and extensive discussion between analysts and decision-makers are needed.

USE OF EPIDEMIOLOGIC INFORMATION IN OZONE-RELATED RISK AND BENEFITS ASSESSMENT

Because death is inevitable, the question relevant to policy-makers, risk assessors, and the public is whether people who live in areas more highly exposed to air pollution experience death at an earlier age than people who live in communities that are less exposed. The effect of pollution on death may be quantified in terms of number of deaths in a given period or life months (or years) lost.

Number of Deaths

The "acute-effect studies" directly estimate the change in daily death rates due to short-term exposure (or a short pulse of exposure). Most models investigate the change in death rates for only one or a few days, whereas distributed-lag models look further ahead to capture delayed acute effects, often referred to as subacute effects. The expansions are useful in understanding the statistical distribution of time between an exposure pulse and the time pattern of occurrence of death. On a population level, it is extremely unlikely that all ozone-related deaths occur either immediately or within 1-2 d of exposure. A more plausible model assumes that susceptibility to death (or frailty) and the success of intervention strategies follow a distribution wherein some people die immediately or within 1-2 d and others first suffer acute health ailments (such as an MI or pneumonia) and their deaths follow a period of unsuccessful treatment or as a result of the decompensation of defense mechanisms. The most complete assessment of the association between acute exposure and death originates in distributed-lag models that integrate the distribution of the time between exposure and death. However, both the usual time-series model and the distributed-lag models focus on a short window of exposure. Effects of long-term cumulated exposure are, by design, ignored in such models.

In contrast, cohort studies focus on the associations between some metric of *long-term exposure* and death rates (as in the ACS study and the Harvard Six Cities Study). As a consequence, the increases in death rates associated with pollution may reflect the total effect of pollution on death, including life time lost because of acute or subacute effects and because of chronic conditions. For example, if pollution increases systemic inflammation, it may contribute to remodeling of airways and to obstruction or contribute to the development of atherosclerosis, and these chronic ailments increase the risk of premature death independently of any acute effects of pollution. One may argue that the current way of reconstructing long-term exposure in cohort studies is insensitive to short-term peaks and changes in exposure and that cohort studies may therefore miss some acute effects. Although cohort studies are likely to capture a large fraction of the effects of pollution on death—including most acute, chronic, and combined effects—some fraction of the immediate acute effects may not be contained in the effect signal, so the total effect of ozone might be larger than that observed in cohort studies.

Years of Life Lost

If one assumes that baseline probability of death or hazard is common to all members of the population at any given age, one can treat estimates of pollution-related risk based on time-series mortality studies in a manner similar to estimates obtained from cohort studies (Miller and Armstrong 2001; Burnett et al. 2003; Rabl 2006). That is, the risk estimates from time-series studies can be used to estimate the life years lost because of acute effects of pollution in the same way that risk estimates from cohort studies are used to derive the time lost because of the chronic, if not total, effects of pollution. However, there is probably a wide distribution of likelihood of death in any population on any given day. Current ambient ozone concentrations probably will not increase risk to a point at which death is likely for most of the population, and only people in a frail state (for example, having a chronic disease) would clearly be at risk of death from ozone exposure.

It is assumed that large portions of the population are at risk of death because of longer-term exposure to air pollution, so it is more reasonable to apply a common baseline hazard function within the cohort design. Relative risks from the cohort studies are then combined with common baseline hazard functions, in the form of life tables, for selected populations to estimate the amount of life lost because of exposure to air pollution (Coyle et al. 2003; Miller and Hurley 2003). The same life-table calculations can be used to estimate the number of people who will die over a fixed period because of changes in air pollution exposure.

If the hazard and risk distributions in the population are not independent, estimates of excess deaths and life years lost in the entire population because of changes in ozone exposure based on time-series studies will not equal the average of estimates across all risk groups.[5] The size of the difference is not known but could be estimated if risks for susceptible groups are obtained (Goldberg et al. 2005).

Translation to Risk and Benefits Assessment

Acute and chronic effects of urban air pollution, such as PM, on mortality have been well established. Risks based on cohort studies are some 10 times larger than those based on acute-effects studies that investigated only the influence of yesterday's concentrations of $PM_{2.5}$ on today's death rates and some 3-5 times larger than the total acute effects, including death, distributed over several weeks.

More recent risk assessments provided separate estimates of acute effects that occur within a few days of exposure and total effects. Such separation may be useful, given that the time between pollution abatement and health benefits may follow a distribution, with acute effects being reduced more quickly than effects due to chronic conditions (Röösli et al. 2005).

Ozone-mortality risk assessment faces major challenges. As summarized above, evidence of long-term effects of ozone on mortality (or survival time) is presently weak, so the derivation of a relative risk to describe the association between ozone exposure and death is more difficult than in the case of PM. The translation of the results into number of attributable deaths is possible but leaves us with two main challenges. First, the assessment of acute effects would be incomplete if based solely on the usual time-series studies, given that subacute (delayed) effects are not captured with these studies; only one study has published estimates for distributed-lag models. Second, in the absence of abundant quantifiable evidence of chronic effects of ozone on mortality, total life time lost because of both acute and chronic effects cannot be estimated. The current evidence from cohort studies, although it is weak, supports the notion that the estimates of the risk of death due to ozone may be largely underestimated if based

---

[5]In this context, the distribution of hazards among the population is the likelihood of death at any age; the distribution of risk is the association between ozone and death at any age.

solely on the main time-series studies that were at the center of EPA's request for this report.

While the current lack of a sufficient number of ozone cohort studies precludes a respective assessment, lessons learned from PM would suggest that estimations based on time-series study alone are far smaller than the overall long-term cumulated effects.

The question of how to translate epidemiologic information into risk and benefit assessment is the subject of debates. Many risk assessments derive, as one intermediate step, the number of deaths attributable to air pollution on an annual basis. That approach has several limitations that are of concern particularly for chronic effects but conceptually also apply to acute effects, as is the case for the acute effects of ozone on death. Although it is true that pollution increases the mortality risk and thus the number of deaths, one has to emphasize that death is inevitable and that improvement in air quality only postpones death and does not prevent it. The concept of "attributable cases" is often interpreted as the corollary of "preventable cases," which in the matter of death is certainly wrong.

Another problem faced with "attributable death" originates in the fact that changes in death rates in any population lead to changes in the age structure of the population. That is, a reduction in air pollution will lead to longer life expectancy and increase the number of elderly people. Age-adjusted death rates are expected to decrease under cleaner conditions, but the absolute number of deaths will steadily increase as the population ages. As a consequence, the "attributable deaths" (or benefits) will not be the same throughout the years after an improvement in air quality. Instead, there will be a complex interplay between the distribution of acute vs chronic causes of premature death in which reduction in death rates caused by chronic exposures will take longer to materialize than reduction in death rates from acute effects and longer than the changes in the population age structure due to the removal or reduction of a risk factor.

The net result of those dynamics is a continual adjustment of the expected annual change in the number of deaths attributable to a change in pollution. As shown by Miller and Hurley (2006), although the benefit (in absolute number of deaths) may steadily increase during a few years, it will later decrease for a couple of years and ultimately reach a time when the absolute number of deaths (per year) is larger than it was in the more polluted condition. Therefore, expression of benefits due to reduction in mortality rates can ultimately not be correctly expressed as a stable annual number of deaths prevented.

For all those reasons, risk assessors omit the expression of "attributable cases of death" but prefer to derive years of life lost (YLL) due to some policy change. Although the derivation of YLL is based on the same input information—the risk changes derived from epidemiologic studies—the use of YLL is more appealing because one can derive the total YLL for the entire remaining life time of any cohort, whether dynamic or static, and express benefits in terms of annual YLL (Brunekreef et al. 2007). However, even though it may be more appealing to express mortality-risk change in YLL, the concept may be more problematic for valuation purposes than that of "attributable deaths" (see Chapter 5).

Another potential approach to the translation of epidemiologic information into risk and benefit assessment is to stop short of calculating annual numbers of deaths prevented and simply report the change in the annual mortality rate expected as a result of the change in ozone. Rabl (2006) shows the mathematical relationships between the change in mortality rate, the annual reduction in number of deaths, and the years of life saved. For an acute mortality effect as a result of short-term (daily) pollution-exposure changes and a stable population (in which births equal deaths), he shows that a permanent reduction in pollution temporarily reduces the number of deaths over some period because people who would have died from pollution live longer. Eventually, however, they die from something else, so the annual number of deaths returns to its previous level. However, there is a permanent increase in life expectancy, a larger population, and thus a permanently lower mortality rate. The valuation implications of this are discussed in Chapter 5.

CONCLUSIONS AND RECOMMENDATIONS

Overall Conclusions and Recommendations

Human chamber and toxicologic studies have yielded strong evidence indicating that short-term exposure to ozone can exacerbate lung conditions, causing illness and hospitalization, and potentially lead to death. Although it is less abundant, the available evidence on ozone exposure and exacerbation of heart conditions points to another area of concern. Panel and epidemiologic studies have also found that exposure to ozone (as an indicator of a broad mix of photochemical oxidants) has those effects. The committee found that the four recent time-series analyses and meta analyses of the relationship between exposure to ozone (and other photochemical oxidants) and premature mortality add to the evidence by providing robust statistical evidence of an association.

Overall Conclusions

On the basis of the broader available evidence and the additional insights obtained from its review of the new time-series studies, the committee concludes that short-term exposure to ambient ozone and the larger photochemical-oxidant mixture is likely to contribute to premature deaths. Although it is rarely possible to exclude the possibility of zero effect in such analyses, the committee concludes that an absence of any effect is highly unlikely. Despite some continuing questions about the evidence, the committee concludes that it is strong enough to be used in the estimation of the expected benefits of reductions in population mortality risk that would result from reducing exposure to ozone and/or the photochemical-oxidant mixture.

Overall Recommendation

The committee recommends that ozone-related mortality, estimated on the basis of the results of the recent time-series analyses, be included in future estimates of health benefits of reducing exposure to ambient ozone. The committee further recommends that the greatest emphasis be placed on estimates based on systematic new multicity analyses using national databases of air pollution and mortality, such as was done in the National Morbidity, Mortality, and Air Pollution Study database, without excluding consideration of meta-analyses of previously published studies. Emphasis should also be placed on risk estimates obtained from analyzing data on multiple days so as to include delayed acute effects. Such health-benefits estimates should be accompanied by a broad array of analyses of uncertainty, while at the same time understanding that a zero value is highly unlikely.

Strengths and Limitations of the Evidence

In reaching its overall conclusions and recommendation, the committee identified a number of strengths and limitations of the evidence.

On the basis of available evidence, the committee believes that deaths related to exposure to ozone (and other photochemical oxidants) are not likely to be restricted to people who are at very high risk of death within a few days. The evidence is provided by the recent analysis of time series in several U.S. cities that focused specifically on a mortality-displacement pattern in the time course of exposure and death (Zanobetti and Schwartz 2008). In that analysis, it was clear that short-term mortality displacement (or 'harvesting') could not fully explain the observed increase in death. The estimates steadily decreased with increasing lags, reaching the level of no effect approximately seven days after exposure but without reaching negative associations at any of these days as one would expected if all events were explained by short-term displacement among the pool of extremely sick and frail people. As discussed by Rabl (2006), if pollution affects both the extremely frail and others, the portion related to displacement cannot be quantified as one can only observe the net difference between short-term harvesting and all other acute effects. The pattern observed in Zanobetti and Schwartz (2008) indicates that at no lag until the disappearance of any effect (day 7) was the portion explained by mortality displacement larger than the rest. This evidence, albeit from only one study provides evidence that ozone-mortality effects are associated not just with those already near to death.

In contrast with the PM literature, very few data are available from the use of distributed-lag models. Those available suggest that in the case of either ozone or PM, effect estimates steadily increase with increasing time of the investigated effects. Specifically, subacute (longer-term) effects that combine effects of several days or weeks of exposure are larger than immediate short-term effects, whereas estimates of effects from cohort studies are the largest. However, the increase in estimates is far larger for PM than for ozone. That may be a consequence of attenuation due to large errors in characterizing exposure to ozone, or it may reflect a more dominant role of acute pathologic conditions related to ozone exposure whereas the role of chronic pathologic conditions is more important in the effects of PM. As these interpretations rely on very few studies, further confirmation is warranted.

In addition to the specific analyses in the time-series studies, there is only very weak, evidence from cohort studies of an association of premature mortality with longer-term exposure. Thus, the committee could not conclude at this time that exposure to long term ozone concentrations is related to mortality.

Although the associations in the recent time-series studies appear sufficiently robust to provide a basis for estimating benefits, several factors need to be considered in interpretation:

- The committee found that short-term ozone exposure is likely to contribute to premature mortality in addition to the risks posed by weather and PM, but studies have not been sufficient to control fully for potential confounding by or interactions with condensed-phase constituents of airborne PM, such as sulfates, acids, carbon, and other elements.
- Controlling for such confounding is further complicated by the data from personal-exposure studies, which have found low correlations of monitored ambient ozone with personal ozone exposure.

Detailed Conclusions and Recommendations

In addition to its conclusions about the strengths and limitations of the new time-series evidence, the committee identified several kinds of additional needed research.

Conclusion: On the basis of available evidence, the committee believes that deaths related to exposure to ozone (and other photochemical oxidants) are not likely to be restricted to those in people who are at very high risk of death within a few days. Because the evidence is based on results from only one study, it warrants confirmation by other studies.

Recommendation: EPA and the scientific community should conduct additional studies to investigate short-term mortality displacement and include the use of alternative methods. An example of such methods is investigation of people who have diseases, such as diabetes and heart disease, which are known to induce high mortality risk associated with air pollution.

Conclusion: If further confirmed, the weak current evidence from cohort studies of an association of premature mortality with longer-term exposure would tend to support the assumption that not all the effects seen in time-series studies are due to short-term mortality displacement.

Recommendation: The committee recommends that EPA, the National Institutes of Health (NIH), and the scientific community conduct further work with cohort studies to examine the association between long-term ozone exposure and mortality. The use of very large cohorts with long followup periods may be required because the long-term ozone-exposure mortality risk appears to be much smaller than that posed by fine PM. This may require pooling information from several cohorts to obtain sufficient statistical power. Long-term ozone-exposure models also need to be further developed to distinguish between variations in exposure not only on the between-cities and within-cities scales but at the individual level.

Recommendation: To the extent that new cohort evidence is strengthened, EPA should consider including estimates based on that evidence in its benefits assessments.

Conclusion: Epidemiologic studies have found that ozone-mediated risks were greater for cardiovascular mortality than for total mortality. However, there have been only a few toxicologic or human research studies directly evaluating cardiovascular effects of ozone exposure.

Recommendation: EPA, NIH, and the scientific community should conduct further studies on the cardiovascular effects of ozone exposure, both in human and animal models. Studies should be designed to identify genetic susceptibility factors.

Conclusion: Although the committee found that short-term ozone exposure is likely to contribute to premature mortality, the optimal way to quantify the effect is unavailable. The method that EPA uses now is appropriate only in very restricted situations that are not likely to be realistic.

Recommendation: EPA should study emerging concepts and evaluate their use and implications in benefits assessments, including relationships between changes in mortality rates, annual deaths prevented, and years of life saved. The alternative approaches for expressing ozone mortality effects will lead to rather similar results if one is supposed to express only the most immediate (acute) effects of pollution. However, with the integration of subacute effects estimates and in particular in the case of use of long-term chronic-effect estimates, the discrepancies between the approaches increase, and the conceptual flaws of the "attributable-cases" model become more pronounced.

Conclusion: Expected differences in susceptibility causes substantial uncertainty in estimating mortality risks on the basis of results of epidemiologic studies of total populations because of the lack of independence between pollution-related risk and baseline hazard.

Recommendation: EPA and NIH should encourage more studies on potentially susceptible population groups, and EPA should explore the effects of uncertainty on risk assessments. To the extent that data are not available, models and assumptions can be used for sensitivity analysis.

Conclusion: Although the recent time-series studies have, to the extent possible, included analyses of alternative ozone metrics (such as 1-h maximum and 8-h maximum), it is important to further examine the relationship between those averaging times and to identify additional ozone metrics based on person exposures. That is important both to examine dose-response relationships, test potential confounding more effectively and to inform future regulatory choices among actions that might have differential effects on peak and multi-hour averages.

Recommendation: EPA and the scientific community should design and conduct studies that use the full array of potential ozone metrics, including those that estimate population personal exposures.

Conclusion: Control for confounding has not sufficiently accounted for geographic and seasonal variability in the relationship between ozone and is possible confounders. Further, data on PM speciation have not been sufficient to include in analyses of potential confounding; however such data are increasingly available from the Speciation Trends Network.

Recommendation: EPA and the scientific community should account for seasonal and geographic variability in the relation between ozone and its confounders and should increasingly include the growing STN database in all future analyses of potential confounding of the ozone associations. To this end, EPA (through its PM Centers program), the Health Effects Institute (through its National Particle Component Toxicity Initiative [or NPACT]), and others (such as the Electric Power Research Institute) are supporting major new efforts in both time series and cohort epidemiology, to address this need. Further EPA should work with the scientific community to ensure that STN collects data frequently enough on the particle components most relevant to the potential for confounding.

Conclusion: On the basis of its review of the evidence, the committee concludes that the association between short-term ozone changes in ozone concentrations and mortality is generally linear throughout most of the concentration range, although uncertainties make it difficult to determine whether there is a threshold for the association for the association at the lower end of the range. If there is a threshold, it is probably at a concentration below the current ambient air quality standards.

Recommendation: EPA and the scientific community should further explore how individual thresholds may vary and the extent to which thresholds depend on the frailty of the individual at any given moment. The research should involve panel studies of individuals considered to be susceptible to premature death from ozone exposure, such as those with impaired lung function.

Because it is not clear whether there is an association between ozone and mortality in the cooler months, warmer months should be examined separately. A sensitivity analysis should be conducted on concentration-response relationships to capture more fully the uncertainties contributed by reliance on average fixed-site monitoring data to estimate population-average personal exposure.

Conclusion: There is a lack of an observed association between ozone and mortality during the periods when exposure to ozone is expected to be low, and the ability to understand the lack is inhibited in part by the fact that monitoring is more limited during the winter periods. Better understanding of ozone-mortality relationships during winter is important for full exploration of (1) seasonal differences in risk, (2) how these seasonal risk differences vary spatially between communities with warmer and cooler winters, and (3) ozone-mortality relationships at lower ozone concentrations.

Recommendation: EPA and the states should extend operation of ozone monitoring into winter and report the measurements.

Conclusion: As has been the case with PM, ozone analyses conducted with distributed-lag models over several days appear to capture overall effects better, but there have been relatively few of them.

Recommendation: EPA and the scientific community should seek out appropriate databases and conduct distributed-lag analyses as part of future epidemiologic investigations to understand the statistical distribution of time between an "exposure pulse" and the time pattern of occurrence of death.

Conclusion: Uncertainty in the epidemiologic models is likely to introduce substantial uncertainty into ozone-mortality analyses.

Recommendations: The committee identified several approaches to addressing the uncertainty:

- Results of the models should be presented with discussion of their reliability and of the estimated uncertainty about which model (if any) is reasonably correct.
- EPA should consider Bayesian approaches, including additional expert elicitation once the recent experience with PM has been evaluated, to uncertainty analysis.
- As EPA develops computational models and input distributions, it should intermittently conduct sensitivity analyses to focus resources on the most important inputs and parts of the models.
- EPA should distinguish between data-derived estimates of some components (such as the concentration-response function) and expert opinions about other components that are lacking in scientific data to achieve a better understanding of how existing data and expert judgment combine to produce estimates and of where new data would be most valuable.
- Time-series studies and meta-analyses should conduct additional sensitivity calculations, in particular to examine sensitivity of results to the structure of the cessation lag and sensitivity of the premature-mortality estimate to the

presence of a potential threshold concentration of ozone below which mortality effects are not observable. EPA should present multiple results from each study and characterize the different sources of error in the studies in a final uncertainty estimation on their benefits analysis.

- Air-quality numeric models (such as CMAQ) should be considered for use in ozone epidemiologic studies to extend the spatial scale of available data. However, the uncertainty associated with the models needs to be carefully considered before inferences are drawn from the simulation models for mortality risk assessment.

5

Economic Valuation of Reduction in Mortality Risk Associated with Ambient Ozone

INTRODUCTION

This chapter examines the conceptual and empirical literature on the economic valuation of reductions in mortality risks generally and specifically in the context of the epidemiologic literature on the linkage between ozone and mortality risks discussed in Chapter 4.

As described in Chapter 2, the U.S. Environmental Protection Agency (EPA) follows standard practice in cost-benefit analyses by using estimates of willingness to pay (WTP) to calculate the economic value of reducing mortality risks. EPA's current approach for estimating the benefits of reducing mortality risk has several steps. First, it estimates the reduction in individuals' annual mortality risk stemming from the expected change in ozone concentration. Next, it calculates the annual number of deaths avoided (that is, postponed to some future year) by the postulated reduction in ozone concentrations to which the population at risk would be exposed. A central estimate of the distribution of economic values of the avoided deaths is then calculated by using a central value of a statistical life (VSL) drawn from the available WTP literature related to changes in annual mortality risk. Uncertainties in the estimates of mortality effects and valuation are then accounted for with reference to the confidence intervals around the estimates.

One important and controversial issue related to valuation is whether the effects of reducing mortality risk should be measured in terms of lives saved (deaths postponed) or years of life extension (increased remaining life expectancy). If it is the latter, then it may be more appropriate to use a monetary value per statistical life year (VSLY) or to adjust the VSL to reflect the preferences of people of different ages rather than to use the same VSL in all cost-benefit analyses. (see Chapter 2 for definitions and examples of VSL and VSLY.) EPA has included various sensitivity analyses (alternative calculations) to explore the implications of alternative approaches in several of its cost-benefit analyses for air-pollution regulations. However, the agency has been using the same VSL estimates regardless of remaining life expectancy in its primary analyses. The Office of Management and Budget (OMB) has urged caution in adjusting VSL on the basis of age because of uncertainties in the literature (Graham 2003) but encourages sensitivity analyses with alternative measures of changes in mortality and its monetary valuation. This chapter examines the theoretical and empirical evidence regarding the relationship between age and WTP for mortality-risk reductions in later sections.

As noted in Chapter 4, another related question is whether those most affected by ozone are already in such poor health that their remaining life expectancy is low despite a reduction in ozone exposure. If decreases in short-term ozone concentrations merely decrease "harvesting" those who are already frail and near death, the economic benefit of reducing their exposure to air pollution may be relatively small and should be reflected in a much lower VSL or VSLY than that estimated for relatively healthy people. In addition, perhaps we should be considering the preferences of those at high risk (generally, those in poor health) regarding reduction in their mortality risk.

This chapter takes up the question of whether the VSL or the VSLY is more appropriate for EPA to use in valuing reductions in mortality risk associated with reduced exposure to ozone. It also considers what economic theory and the empirical literature say about how WTP for reductions in mortality risk varies with the age of the person at risk, the size of the risk change, the health status of those at risk (existing chronic illness vs average health and effects on quality of remaining life), income, and other socioeconomic variables to improve approaches for assigning values to reductions in risk.

CONCEPTUAL UNDERPINNINGS OF VALUATION OF MORTALITY RISK

As introduced in Chapter 2, cost-benefit analysis for a regulatory decision is intended to help answer the question of whether some proposed policy will result in more welfare for the affected people—that is, whether the expected benefits of the regulation are worth the expected costs. The standard approach to cost-benefit analysis is to quantify both costs and benefits in terms of "opportunity cost." For benefits, that means a measure of what those who benefit would be willing to forgo to obtain the specified benefit. This is the basic concept of WTP, which is taken to be a measure of how much better off the members of the benefiting population perceive themselves to be, assuming that the members of the population have full information about the benefits.[1]

---

[1]There is some debate in the literature on the foundations of welfare economics as the whether WTP is an appropriate measure of the welfare changes of individuals. For a recent overview of this debate, see Adler and Posner (2006), especially Chapter 2. WTP as

In the early days of cost-benefit analysis for public programs, the predominant approach to determining the monetary value of programs that reduced mortality risks was the human-capital approach. That approach focused on the financial loss when a person dies prematurely, as measured by lost lifetime earnings (see, for example, Dublin and Lotka 1930). It is conceptually similar to the cost-of-illness approach, which measures lost earnings and medical costs due to illness and premature death. Later, several economists (for example, Schelling 1968 and Mishan 1971) discussed why those indicators of the financial burden of disease and premature death do not measure the monetary value of the full effect of disease and premature death on the welfare of the population and are therefore insufficient for a full cost-benefit analysis of public policies aimed at reducing morbidity or mortality. They noted that those financial measures ascribe no value to the lives of homemakers or retired persons, and they argued that a more appropriate way to value a program that reduces mortality risk is to determine what the reduction in risk is worth to the people who benefit. Thus, WTP is a more appropriate measure of the change in welfare in cost-benefit analysis because it reflects not just the financial effect but the value that people place on the effect on quality of life and longevity.

WTP specifically measures the ex ante value of reduction in risk that public-health and public-safety programs may provide. It is not known specifically whose death will be prevented, only that the probability of death in a group of people will be reduced. Generally, WTP is expected to exceed the cost of illness, but costs paid by third parties (such as insurance-paid medical costs) will generally not be included in an individual's WTP and should in some cases be added to individuals' WTP to determine total value. There is substantial evidence that WTP for reductions in mortality risk typically exceed by a substantial amount the expected value of lost earnings (e.g., Viscusi 1993).

Some authors have argued that the appropriate measure of WTP for risk reduction is not the values expressed by the people actually affected by the policy, but the values of people who do not know their actual position in society (age, income, wealth, health status, etc.) but are acting behind a "veil of ignorance" or as "impartial spectators" (see, for example, Pratt and Zeckhauser 1996; Sunstein 2002; Hammitt 2007, p. 237). There is an extensive literature in welfare economics based on these ideas as first developed by Harsanyi (1955) and Rawls (1971). But with one significant exception, this literature has focused on the principles of distributive justice and fairness and not on the valuation

---

a welfare measure is criticized as being dependent on the existing distribution of income and wealth and on existing preferences of individuals, which may be influenced by such things as addictive behaviors. The standard approach in cost-benefit analysis is to acknowledge these difficulties in principle but to base empirical measures of value on WTP whenever possible. In fact, the OMB *Circular A-4* is explicit in calling for monetary measures based on WTP (OMB 2003).

of specific goods received by individuals. The exception is the paper by Pratt and Zeckhauser (1996). We see this literature as offering an interesting thought experiment for considering ethical issues regarding distributive justice and valuation. But it does not offer practical methods and techniques for the development of criteria of distributive justice or for the empirical estimation of values. All of the practical methods for estimating values are necessarily based on the preferences of individuals who know their positions in society, and these values are based on expressions of WTP for changes in risk.

A possible surrogate for the "veil of ignorance" or "impartial observer" perspective is the "public preference" or "social value" approach to determining benefits. This approach involves asking people to express their WTP for policies that would reduce the risk of premature mortality for specified groups of individuals with certain characteristics, for example, age. Some studies have shown that people have different preferences concerning policies delivering health improvements or risk reductions to different social groups (Cropper et al. 1994; Dolan et al. 2005). But such studies are not based on the "veil of ignorance" approach, because it is not possible to show that people could in fact ignore their own position in society relative to those whose risk changes they were being asked about. For either approach, the conceptual questions regarding appropriate aggregation of such values for a specific policy assessment have not been resolved (see below).

Basing the values used in economic analysis on the preferences of the affected individuals is a fundamental tenet in welfare economics. Thus, the monetary measure of the benefit of a reduction in mortality risk is based on the value placed on the benefit by the individuals receiving it. It is not the value held by the policy-maker or the experts. The standard approach for cost-benefit analyses is thus to measure the benefits of a mortality-risk reduction on the basis of the value to each individual of his or her own reduction in risk and then to sum these values across all affected individuals.

An important conceptual question is whether, if individuals are altruistic toward others, especially family and friends, altruistic feelings should be included in the valuation of reductions in mortality risk by adding WTP values that people have for others' risk reductions. As was discussed in Chapter 2, economic analysis (e.g., Jones-Lee 1991) shows that if the altruism involves a general regard for the well-being of others, that is, is non-paternalistic in nature, summing WTP over all affected individuals would leave the sign of net social benefits unchanged since the altruists' benefits would be offset by the altruists' recognition of the costs of the policy to others. Summation is appropriate only if the altruism is paternalistic in nature. Because we have no evidence to indicate the extent of paternalistic altruism regarding risk reductions, we consider that the approach of using individuals' WTP only for their own mortality-risk reduction is appropriate.

According to this approach, maximum WTP for mortality-risk reduction is defined as the amount of money that would make the person indifferent between choosing to spend the money to decrease his or her own mortality risk and forgoing the decrease in risk and keeping the money to spend on other things. Expressions for a person's WTP can be derived from models of the choices made by utility-maximizing people (see Chapter 10 of Freeman [2003] for an overview). The models are simplified representations to describe economic behavior and to predict the choices that a utility-maximizing person would make. They provide a formal definition of WTP and provide a basis for analyzing the factors that determine a person's WTP and how it may vary. For example, the model can help to analyze the question of whether VSL varies with age, income, and other socioeconomic characteristics of individuals.

The life-cycle consumption model (see Shepard and Zeckhauser 1982; Cropper and Sussman 1990) is the most comprehensive economic model that is available for analyzing how WTP for small changes in mortality risk is expected to be related to a person's age and other factors. In the model, persons, acting as individuals, are assumed to try to maximize their expected "utility from consumption" over their lifetimes, given constraints involving their initial wealth, their annual income, the cost of consumption, and a subjective rate of time preference. Utility from consumption is the satisfaction or enjoyment obtained from any activity that involves an expenditure of money or time. The subjective rate of time preference is the rate at which a person discounts future utility relative to current utility. Empirical evidence and common sense suggest that the value to a person today of future consumption is something less then the value of current consumption, and the difference between the two is the discount factor (which implies a subjective rate of time preference).

The model postulates that a person makes economic choices on the basis of his or her expectations about future utility from future consumption and about future income, all discounted to a present value at the person's rate of time preference. People are also assumed to include in their choices their expectations about their future survival probability. The model incorporates constraints that people have because of initial wealth and income over time, and it takes opportunities for borrowing, bequests, and lending into account.

In a simple model, an expression of a person's maximum WTP for a small reduction in mortality risk can be derived to illuminate the factors expected to influence the magnitude of and variation in WTP. The expression is the product of the reciprocal of the probability of surviving the current period, the present discounted value of the expected utility from consumption conditional on surviving the period, and the reciprocal of the marginal utility of consumption conditional on survival (used to convert utility into monetary terms). Marginal utility of consumption is the additional satisfaction or enjoyment obtained from a small increment of consumption.

To examine the effect of age on WTP for a reduction in the current risk of death, we ask how age would affect each of those terms.[2] As people age, their probability of surviving the current period generally falls. Thus, their WTP to reduce their mortality risk in the current period is expected to increase because as the probability of death in the current period rises, a person is likely to be willing to forgo more current consumption to increase the chances of surviving to the next period to continue enjoying the fruits of his or her resources. That may be offset, or at least limited, by a desire to leave some bequest for one's children and others. However, aging not only increases current risk, it decreases life expectancy conditional on surviving the current period, and this is expected to decrease WTP. This effect could easily offset the first effect on WTP.

How the present value of expected utility of consumption—the second term of the equation—changes with age is ambiguous. If per-period consumption and its utility were constant over time, the present value of expected utility of consumption would be proportional to discounted remaining life expectancy. With rare exceptions, mortality rates rise with age. Thus, the decrease in remaining life expectancy with increasing age motivates the hypothesis that WTP for mortality-risk reduction should fall with age. However, if there are constraints on borrowing against future income, per-period consumption is likely to rise in the earlier years of the life cycle. If, for example, a person cannot be a net borrower but can lend at the riskless rate of interest, his or her consumption is likely to be constrained by limited income at the beginning of life due to lower earning potential. That will cause the present value of the utility of consumption and hence the WTP for mortality-risk reduction to increase up to some point and then to decline (Shepard and Zeckhauser 1982); a plot of WTP against age would have an inverted U-shape.

The combined effect of all those factors is that the impact of age on WTP for mortality-risk reduction is ambiguous on the basis of theoretical analysis alone. It requires empirical evidence to determine whether the combined effect of these factors is that WTP increases, decreases, or remains the same over all or part of a lifetime.

Similar arguments can be made about health status. Health status may affect life expectancy, pleasure in consumption, or both. If life expectancy is reduced because of illness, the first term in the expression of WTP should be positive; with smaller chances of surviving the current period, the person should be willing to pay more to reduce mortality risk. As with age, the effect of health on the other factors that determine WTP is ambiguous. Chronic illness, for example, could reduce the quality of life and cause a person to derive less enjoyment from consumption.

In summary, the effects of age and health status on WTP are ambiguous. Under reasonable assumptions the discounted expected utility of consumption

---

[2]See Hammitt (2007) for further discussion of the points raised in the remainder of this section.

increases with future health and life expectancy (and so decreases with age). The ambiguity arises because the expected marginal utility of consumption is also likely to increase with future health and life expectancy. Since VSL is the ratio of the discounted expected utility of consumption to the expected marginal utility of consumption, and since both terms move in the same direction, the effect on the ratio is indeterminate. Thus, the combined effect of lower likelihood of survival and shorter remaining life expectancy on WTP to reduce mortality risk is ambiguous.

So far, we have described only the effects of a current change in risk of death on WTP (or VSL). Also potentially relevant to valuing the effect of ozone on mortality risk is the WTP (or VSL) associated with a *future* change in risk of death. A lag between a change in exposure and a change in risk is often referred to as a latency in the risk. To address it, the life-cycle model can be further manipulated (see Alberini et al. 2006a) to show that the WTP for risk reduction in some future period is likely to be lower than the WTP for risk reduction in the current period for two reasons. First, there is a probability that the person will not survive to the age when the risk reduction is to be realized. And second, if the rate of time preference is positive, the prospect of future consumption has less current utility than present consumption. However, if WTP for some risk reduction increases with age faster than the discount rate, the current WTP for a future risk reduction would be larger than for a current risk reduction (Hammitt and Liu 2004).

The models can provide insights into other key factors that affect WTP. WTP for a given risk reduction should be larger than WTP for a smaller risk reduction. However, the VSL is arithmetically the WTP divided by the risk reduction, so that simple relationship is not enough to predict whether the VSL will be lower (or higher) for a smaller risk reduction. The theoretical literature (Hammitt and Graham 1999) suggests that for small changes in mortality risk, WTP will move roughly in proportion to the size of the risk change, leaving the VSL roughly constant over different risk changes. However, empirical studies (see Hammitt and Graham [1999] for a summary and Alberini et al. [2004] for a more recent example) often find that the VSL is larger for smaller risk changes.

The expected effect of greater income or greater wealth on WTP (and therefore the VSL) is unambiguous. Higher income (or greater wealth) is expected to increase WTP for a given risk reduction. No life-cycle models in the literature distinguish effects on WTP on the basis of sex, race, marital status, number of children, or other variables that one might hypothesize could affect WTP. However, the empirical literature has looked at some of those issues (see below).

EMPIRICAL METHODS OF VALUING MORTALITY-RISK REDUCTIONS

WTP is typically measured by analyzing the prices paid for goods and services. The maximum price that a person is willing to pay for a good or service is a measure of how much the person values the good or service. Prices for reducing or preventing mortality risks cannot be directly observed, because, for the most part, reduction or prevention of mortality risks is not directly purchased in the marketplace. However, there are instances in which the monetary tradeoffs that people are willing to make between income and mortality risks can be observed or measured, for example, higher wages for riskier jobs or higher prices for safer products.

There are two general economic approaches to measuring WTP for nonmarket goods, such as changes in mortality risks. The first is to analyze actual situations in which WTP for changes in mortality risks may be indirectly revealed; this category of estimation approaches, which is based on behavior, is called revealed preference. The second is to have subjects respond to a hypothetical situation that is designed to have them reveal their WTP; this category, which is based on responses to survey questions, is called stated preference.[3]

Revealed-Preference Methods

An example of revealed-preference methods is a wage-risk study in which wage premiums for risks of death on the job are estimated. This approach analyzes the factors that determine differences in actual wages between jobs, including on-the-job risks of death. The additional wages that people are paid per unit of additional risk of fatal injury is a measure of the monetary value of the additional risk to the person who voluntarily accepts the risk in exchange for a given wage increment. In this context, this value is equal to or less than his willingness to accept compensation (WTA) for incurring greater risk and is equal to or greater than his WTP for a reduction in risk (Hammitt, 2002).[4] The primary advantage of this type of study is that it is based on actual behavior. Some of the limitations are that it is difficult to find situations in which an observable tradeoff is made between income (or expenditures) and mortality risk and difficult statistically to isolate WTP for a mortality-risk increment from other factors involved in the specific behavior or decision. Fully specifying potential confounding factors in these kinds of decisions, such as associated decreases in rates of nonfatal injury or other benefits (such as reduced risk of property loss due to purchase of a home smoke detector), is one of the important challenges to

---

[3]For more information on these estimation techniques, see Freeman (2003).

[4]When WTP and WTA are small compared with income, the theoretical expectation is that these measures will not be much different from one another for a given change in risk. In revealed-preference studies, that is generally the case. However, in stated-preference studies, responses to WTA questions are often much higher than responses to WTP questions. Concern that respondents are not forced to consider a budget constraint when answering WTA questions has led most analysts to prefer WTP questions in stated-preference studies.

the estimation of WTP for mortality-risk reduction in revealed-preference studies.

Many labor-market studies have examined the tradeoff between fatality risk and income; they are referred to here as wage-risk studies. The estimation method is important because it has been a primary source of the estimates that EPA has used for monetary valuation of mortality-risk reductions in environmental-policy analyses. The basic premise of wage-risk studies is that workers reveal the tradeoffs that they are willing to make between risk and income in the choices of jobs that they accept. The idea is simple, but the execution of a study is complex because many factors go into a job choice and a hiring decision. The safest jobs tend to be the highest-paying jobs because they require more skills and education. Other factors must therefore be taken into account, and this is done through estimation of a wage equation in which wage rates are specified as a function of worker characteristics, job and industry characteristics, and on-the-job fatality risks.

Wage-risk studies do not directly estimate a VSL but rather a market equilibrium between wages and on-the-job risk that represents the intersection of labor supply (and the worker's marginal demand for an increment in on-the-job safety) and labor demand (and the employers' marginal cost to supply more safety). The result is an average value of a small change in risk of on-the-job fatality. The VSL is inferred from that "risk premium" as the total of the incremental wages paid to each worker per year to prevent one fatality in that year.

With a few exceptions (Gegax et al. 1991; Liu and Hammitt 1999; Hammitt and Ibarraran 2006; who asked workers their perceptions of on-the-job risks), wage-risk studies presume that workers' perceptions about risks on the job can be accurately measured by reported risks. Further, they presume that workers have sufficient information about on-the-job risks for their choices in the labor market to reflect responses to actual differences in risks across jobs. Using the results of those studies in policy analysis also presumes that the marginal value of risk reduction in the labor market is consistent with such values in the general population. For that reason, wage-risk studies that look at a broad cross-section of occupations and industries are preferred to studies that look at a narrower set of occupations and industries if the results are going to be used in policy analysis for the general population. Even so, if there is self-selection of high-risk occupations by those who are less risk averse, implying they demand less compensation than those in the general population do, then the wage-risk studies will underestimate the general-population VSL.

Wage-risk studies are limited to examining the types of on-the-job fatalities that are recorded in available databases that cover a wide array of industries (for example, databases of the Bureau of Labor Statistics, worker compensation boards, and the National Institute for Occupational Safety and Health); these are primarily accidental deaths. Indeed, a large share of on-the-job fatalities occur while being in vehicles. Other causes of death that may be related to work, such as cancer and chronic illness related to on-the-job exposure, are more difficult to study and less likely to be included in available databases because of long lags between exposure and death and because they are thought to be less frequent in most occupations.

Stated-Preference Methods

An application of stated-preference methods is a survey in which subjects are presented with a hypothetical situation that involves a tradeoff between income or expenditures and a change in the risk of death. In a direct stated-preference approach, subjects are asked to estimate what they would be willing to pay to change their mortality risk by a specific amount. Other stated-preference approaches ask respondents to choose among alternative scenarios or policies that have varied costs and amounts of risk reduction and thus reveal but not explicitly state their WTP values. In any stated-preference approach, it is important that the situation presented to study subjects be realistic and easy to understand. The primary concern about this type of study is whether subjects are able and have incentive to give accurate responses to questions of this nature and whether answers in the context of hypothetical situations are accurate predictors of behavior and choices in situations in which the costs and consequences are real.

Many challenges exist concerning survey design for stated-preference studies about mortality risk. Difficulties in comprehension of small risks are often discussed. Corso et al. (2001) demonstrated the importance of visual aids in communicating quantitative risk information. Krupnick et al. (2002) used some simple comprehension questions to identify respondents who might be unable to understand or work with quantitative risk information. It seems that no matter how carefully the information is presented, some fraction of the general public has difficulty in making sense of it or is unwilling to put in the effort required. Another challenge is the WTP elicitation mechanism. Cues given in the survey can lead to starting point bias (Mitchell and Carson 1989) and different procedures can lead to different estimates (Champ and Bishop, 2006). Finally, some studies of illness-related health risks have used health-care changes as the risk-reduction scenario (e.g., Johnson et al. 1998; Krupnick et al. 2002; Alberini et al. 2004), and others have used relocation and cost-of-living changes (e.g., Viscusi et al. 1991). Stated-preference studies generally obtain lower mean VSLs than wage-risk studies (Kochi et al. 2006), but why this occurs is difficult to determine.

Wage-risk and stated-preference studies have different strengths and weaknesses with respect to providing estimates of WTP for mortality-risk reduction for use in environmental-policy analysis. Wage-risk studies have the strength of being based on actual behavior, and many economists consider this an overwhelming advantage over stated-preference methods. However, they include only working-age adults who are healthy enough to be working, and they predominantly consider accidental death—a different population and a different type of risk from what may be associated with ozone.

Stated-preference studies have the advantage of being able to be designed to address any type of mortality risk and can capture values for any population that is capable of answering the survey questions. Thus, there is a greater possibility of specifically addressing the type of illness-related mortality risks that are associated with pollution exposure and of including the older population that is expected to be at greatest risk.

Methods for Estimating Values of Changes in Life Expectancy

The above discussion focused entirely on how revealed- and stated-preference studies estimate WTP for annual mortality-risk reductions, which have generally been summarized by reporting average VSLs. The literature on the estimation of WTP specifically for changes in life expectancy (usually reported as VSLYs) is smaller. In the revealed-preference literature, specifically, the wage-risk studies, variation in the average age of workers in various occupations is used to identify life-expectancy differences across the working population and from that to estimate the wage premium for taking jobs that are riskier in terms of life-expectancy losses. Selection of workers into riskier jobs complicate the use of this information as a proxy for the VSLY in the general population and in a pollution context. Aldy and Viscusi (in press) found that on-the-job fatality rates vary by age and that taking this into account alters the estimation of wage-risk premiums by age group. Of course, since few workers are in the labor force beyond age 65, the VSLYs (and VSLs) estimated based on labor market data can only apply to the "near-elderly" (Evans and Smith 2006). Findings in this literature are reviewed below.

For stated-preference studies to estimate WTP for reductions in risk defined as increases in life expectancy requires making respondents understand that they are valuing an increase in life expectancy over their remaining lifetimes (from a shift in the hazard function), rather than more time being simply tacked on to the end of life. Few have attempted this task, and evidence suggests that these attempts have not been successful (see below).

In practice, most VSLY estimates used to value changes in life expectancy have been ad hoc and derived from estimates of the VSL. The most frequent are those that simply annualize the VSL, assuming a given rate of interest (time preference) and underlying life expectancy (Mauskopf and French 1991). Another approach is to convert the annual risk reduction being valued into its equivalent change in life expectancy and, using that metric together with the estimated WTP, to calculate the VSLY (Alberini et al. 2006b). The drawback of this approach is that were respondents to realize that their annual risk reduction of (say) 5 in 10,000 translates into a few weeks of additional life expectancy, their WTP might be different.

The WTP for any risk reduction can be summarized as an average VSL or an average VSLY, given estimates of the rate of time preference and remaining life expectancy for the population from which the WTP values were derived. As discussed below, the difficulties arise when either of these measures is applied to other populations or other risk reductions, assuming the value is always the same.

ENVIRONMENTAL PROTECTION AGENCY'S CURRENT APPROACH TO VALUING MORTALITY-RISK REDUCTIONS

Overview

Since EPA first began doing Regulatory Impact Analyses using welfare economics rather than a human capital approach to monetizing reductions in mortality risks, the agency has used a uniform VSL, rather than different VSLs applied to different groups or risk contexts, or a VSLY, whether uniform or differentiated. The only exceptions have been in sensitivity analyses exploring alternative valuation estimates as detailed below. The reasons for EPA's approach may be both practical and ethical. The early valuation literature did not support any but the simplest (VSL) approach to this valuation problem, and there was controversy even surrounding attaching monetary values to "life," let alone applying different values to different groups (as would be implied by a life-year approach) or different risk contexts. Furthermore, the time-series epidemiological literature, which was all that was originally available for quantifying mortality risk changes from air pollution changes, could not support a life-years approach, as detailed in Chapter 4.

In addition, EPA focused on the valuation of private rather than public goods, that is, on estimates of individual WTP for reductions in their own risks of death, not for those of society more generally. This choice was (and still is) consistent with the bulk of empirical literature (as addressed below) and the state of welfare economics (as noted above).

Based on these choices, the agency has performed hundreds of RIAs, many arguably influencing agency decisions and even subsequent decisions over legislation in Congress. Furthermore, these choices have been ratified as the best practical options by all of the Science Advisory Board (SAB) panels EPA has asked to examine them, most recently in 2007 (EPA-SAB 2007), as detailed below. Thus, the present committee believes that these choices should not be overturned lightly and without significant evidence, both conceptual and practical, to support a defensible alternative.

Recent Practice[5]

According to Robinson (2007), recently, EPA (2005a, 2007b) has used a

---

[5]The discussion in this section contains text that is excerpted or summarized from a document by L. Robinson (2007). The discussion is also informed by another document by L. Robinson (2004).

central VSL that is a midpoint between results obtained in two meta-analyses of the wage-risk studies (Mrozek and Taylor 2002; Viscusi and Aldy 2003). It is a VSL of $5.5 million in 1999 dollars and at 1990 income levels, which is about $6.7 million in 2006 dollars.[6] EPA (2005a) noted that this VSL estimate is very close to the mean estimate reported in a third meta-analysis (Kochi et al. 2006), which included stated-preference and wage-risk studies. EPA's previous VSL estimates were based largely on work completed in the early 1990s to support its retrospective and prospective analyses of the impacts of the Clean Air Act (CAA) (EPA 1997b, 1999a). Relying upon the work of Viscusi (1993), EPA selected 26 VSL estimates for its analyses; 21 of those estimates were derived from wage-risk studies and 5 from contingent-valuation studies. The mean VSL estimate based on those studies is $7.4 million, with a minimum of $0.9 million and a maximum of $20.9 million (in 2006 dollars). The 21 estimates from the wage-risk studies were scattered throughout that range, but the estimates from the stated-preference studies tended to be in the lower half of it. The mean VSL and distribution based on the review of the 26 studies were incorporated into EPA's *Guidelines for Preparing Economic Analysis* (EPA 2000a).

In assessments of the costs and benefits of the CAA (EPA 1997b, 1999a), EPA conducted a sensitivity analysis by using a constant VSLY and an estimate of life years saved by reductions in mortality related to particulate matter (PM), assuming average remaining life expectancy for each age cohort and using age-specific risk estimates. That reduced the estimated benefit of reductions in PM-related mortality by nearly 50%. In another analysis, EPA (2000b) conducted a different sensitivity analysis by using VSL estimates that varied by age on the basis of empirical studies of WTP for mortality-risk reduction that showed WTP increasing with age up to about the age of 40, then decreasing (Jones-Lee 1989; Jones-Lee et al. 1993). Both sensitivity analyses obtained smaller benefit estimates because a large share of PM-related deaths are of people over 65 y old. At a given VSLY, for example, reducing the risk of death of an older person would have a lower value because fewer life years would be saved. In a sensitivity analysis for regulations addressing emissions from large spark ignition engines (EPA 2002), the agency used a more complicated approach that reflected initial results from the work of Alberini et al. (2004) as well as the adjustment factor from Jones-Lee (1989). In this case, EPA derived different estimates of VSLY for younger and older age groups from selected VSL estimates. The result was a higher VSLY for the older age group. EPA applied these VSLY estimates of

---

[6]One change EPA has made in use of VSLs is adjusting the VSL for expected changes in real income over time (EPA 1999a). This adjustment is for secular changes in income over the economy, not for changes in individual circumstances over their life cycle, which would be already accounted for in an individual's observed or stated preferences for paying for mortality-risk reductions. Specifically, EPA uses an elasticity estimate of 0.4. For 2006 GDP per capita, that implies that the current central estimate of VSL is about $7.6 million in 2006 dollars and at 2006 income levels.

life-years saved for each age group. The net effect was a lower value for risk reduction for the older age group because the smaller remaining life expectancy more than offset the higher VSLY. These sensitivity analyses (particularly the age-adjusted VSL) created considerable controversy at one point when the practice was dubbed the "senior death discount" in the press (see, for example, Seelye and Tierney 2003). As a result, EPA stopped using age-adjusted VSL estimates even in sensitivity analyses.

According to Robinson (2007), EPA is now revising the economic analysis guidelines and updating its approach for its next prospective analysis of the CAA. In an effort related to updating the process for selecting values for mortality-risk reduction for the guidelines, EPA asked its SAB to assess further several questions regarding the selection of estimates for use in valuation of mortality-risk changes. To support the updating, EPA reviewed and summarized recent studies and meta-analyses in the VSL literature (Dockins et al. 2004). It also funded research on the robustness of estimates from wage-risk and contingent-valuation studies (Black et al. 2003; Alberini 2005) and from studies of averting behavior (actions that people take to avoid or mitigate risks, such as the use of seatbelts) (Blomquist 2004). EPA later convened a group of statisticians to address the use of meta-analysis (Allen et al. 2006) and reviewed the literature on the relationship between life expectancy and the VSL (Dockins et al. 2006).

The SAB (EPA-SAB 2007) concluded that meta-analyses are useful for systematic assessment and analysis of factors that affect the results of wage risk and stated-preference studies, including variations in methods, data sources, and study populations. It recommended that EPA consider results of meta-analyses and the context of relevant policy questions to develop selection criteria for studies in the literature that are methodologically sound and applicable to the policy questions. A mean or central tendency of VSL estimates from the studies could be calculated by using any of several weighting methods. The SAB further recommended that EPA not rely exclusively on either revealed-preference or stated-preference studies, but rather give weight to results on the basis of how well they address the policy questions at hand. The SAB panel cautioned against most quantitative adjustments of estimates from individual studies to account for differences in methods or population characteristics other than inflation and currency differences. If adjustments are made, the SAB panel recommended they be based on the results of individual studies rather than the aggregate results of meta-analyses.

The SAB concluded that the relationship between VSL and remaining life expectancy requires empirical study because economic theory places no restriction on whether VSL increases, decreases, or remains constant as life expectancy decreases. It noted that because individual life expectancies can rarely be determined with much accuracy, it is necessary to focus on how VSL varies across a population in relation to age and health status, which are related to life expectancy. It concluded that the available literature is not sufficiently robust to support estimates of VSL that vary with age. It also recommended against using a constant VSLY, stating that "if there is insufficient information to indicate that

VSL declines with age, there is not sufficient information to indicate VSL is strictly proportional to remaining life expectancy." And it recommended that EPA continue to use an age-independent VSL to value mortality-risk reductions but also report the age distribution of lives saved or the number of life years saved and fund more research on valuation of mortality-risk reductions in the context of environmental health risks.

Limitations of EPA's Approach

Concerns and questions about EPA's approach in monetizing mortality-risk reductions in Regulatory Impact Analyses are the following: The WTP to reduce or avoid a unit change in mortality risk

(i) is based on labor market studies that seem ill-suited to a policy context where at risk groups are children and the elderly;
(ii) is uniform over different risk contexts (dreaded risks, voluntary risks, etc.);
(iii) is uniform over different populations (by age, gender, race, income, region);
(iv) counts premature deaths avoided rather than life-years extended;
(v) is for a private good rather than a social good.

These five points merit some individual discussion.

(i) Most of the available empirical WTP estimates are for risks of accidental death in circumstances in which people are voluntarily exposed to risks (for example, in choosing a job or driving a car). VSL estimates drawn from wage-risk studies, the most common source of published VSL estimates, are for working-age adults who are well enough to be employed. The contexts of the available estimates and the contexts of most health-risk changes being evaluated in cost-benefit analyses of environmental regulations differ in some potentially important ways. For example, environmental health risks are related primarily to illness rather than accidents, and the risks tend to be more concentrated in the elderly population than are risks of accidental death. Differences in the characteristics of the people at risk, such as age and health status, may result in differences in the WTP to reduce their risk.

(ii) People's reactions to and attitudes toward risk have been shown in a substantial risk-perception literature to be affected by many attributes beyond the magnitude of the risk. Attributes that appear to be important in how subjects rate different risks include dread or fear related to the risk, the source of the risk, its voluntariness, whether and how well it can be controlled by a subject, and whether the mitigation measures are undertaken privately or as part of a broad government program (Slovic 1987; Cropper and Subramanian 1995). The literature, however, tells us very little about how WTP to reduce mortality risk may vary with any of these risk attributes.

(iii) The objective of an economic-benefits assessment is to develop an estimate of the aggregate WTP for the benefits of a policy that is the sum of the individual WTPs of all the affected people. People's WTP values can differ because of differences in such things as income, wealth, age, health status, sex, race, and baseline risk. To meet the objective of a preference-based aggregate WTP, an analyst conducting a benefits assessment of an air-pollution control policy should, in principle, obtain estimates of WTP from each of the affected people. For ethical and practical reasons, government agencies generally do not adjust the WTP estimates they use to reflect these differences. A practical reason is the lack of knowledge of how WTP varies with individual characteristics. An ethical reason is a policy judgment that differences due to, say, income should not be relevant for policy. Typically, an average WTP for the population is used, making no explicit distinction in WTP across population groups.

For example, economic theory predicts that individual WTP for risk reduction should increase with income. Taking account of that in estimating aggregate benefits would result in assigning higher values to reducing the risks to higher-income people if other items are equal. EPA and others have, for ethical reasons, decided not to take account of differences in income in estimating aggregate WTP and instead use the population mean WTP to calculate the aggregate benefits.

Unobserved population heterogeneity that affects WTP is a more difficult problem. For instance, the wealth of an individual may affect their WTP, irrespective of their income. But wealth is difficult to measure in the best of circumstances and is not measured in WTP studies. If the elderly as a group are relatively wealthier than is apparent from their incomes, then their WTP would be overestimated in valuation studies relative to what it would be if wealth were held constant.

(iv) Age as a population characteristic is singled out for debate based on two factors: (1) concern that deaths from air pollution apply primarily to those who would have died shortly anyway; counting these deaths as loss of much of a lifetime, therefore, seems like a gross overestimate; and (2) people would prefer to prevent the imminent death of a younger than an older person because the former have so many more years to live. The response to these positions is that (1) so-called short-term death displacement (or "harvesting") has been shown to be unlikely to account for most of the mortality associated with ozone (see Chapter 4), and (2) the perspective on life years as a better, or more natural, metric is from a social perspective, not necessarily from the individual's perspective, which is the preferred perspective for valuation in cost-benefit analysis. In any event, EPA, with some recent exceptions, has chosen to use a VSL metric that is not varied for differences in life expectancy rather than a VSLY metric to adjust for differences in life expectancy.

(v) If one were to move to a social perspective on valuation, several problems would need to be overcome. They include a very thin literature to support any specific monetary values and the inability to distinguish paternalistic from non-paternalistic altruism.

EMPIRICAL EVIDENCE ON VALUATION OF MORTALITY-RISK REDUCTIONS

The key conclusion from the theoretical analyses regarding valuation of mortality-risk reduction is that the most important questions for policy analysis cannot be answered without substantial empirical data. For most of those questions, the available empirical evidence is insufficient for supporting robust quantitative conclusions. This section discusses the empirical evidence that is available.

Recent Meta-analyses of Mortality-Risk Valuation

Three meta-analyses of the VSL literature (Mrozek and Taylor 2002; Viscusi and Aldy 2003; Kochi et al. 2006) overlap a great deal in the studies included, but each applied some unique analytic approaches, and there are some differences in how VSL estimates were selected from the literature. All the meta-analyses address VSLs, because the literature is too limited to do such analyses on VSLYs. Two of the analyses, Mrozek and Taylor (2002) and Viscusi and Aldy (2003), are limited to wage-risk studies. Kochi et al. (2006) include wage-risk studies and stated-preference studies and report results for each type of study and combined results. All three meta-analyses report mean VSL estimates.

One of the challenges that complicated all three meta-analyses was that many studies report more than one VSL estimate based on different model specifications for analyzing the data. In addition, multiple publications are sometimes based on the same or similar data sources. Each meta-analysis addressed that issue in a different manner. Viscusi and Aldy selected one estimate from each publication according to what the authors indicated was their preferred result. Mrozek and Taylor included every VSL estimate from each publication but weighted them according to the inverse of the number of estimates from each publication. Kochi et al. conducted statistical tests for homogeneity of subsets of estimates by the same authors and took mean values from subsets that passed the test for homogeneity; using this process, they collapsed 197 VSL estimates (from 45 studies) into 60 mean estimates that were presumed to be independent for the purposes of the meta-analysis.

All the meta-analyses reported results separately for U.S. and non-U.S. studies and adjusted the VSL estimates to the same-year dollars. Neither Mrozek and Taylor nor Kochi et al. compared results on the basis of differences in study population or risk characteristics. Viscusi and Aldy (2003) reported results of cross-study comparisons of the effect of sample average income on VSL results and of within-study analyses of the effect of age. The possibilities of analyzing the effects of population or risk characteristics in the available literature are few because most of the studies are wage-risk studies, which are limited to working-age population and to risks of fatal on-the-job accidents. Only the stated-preference studies cover a wider array of population and risk characteristics, and the small number of such studies reduces the effectiveness of across-study comparisons.

On the face of it, the three meta-analyses report a wide range of mean VSL estimates: around $3-10 million (in 2006 dollars). However, when results are based on the most comparable set of U.S. studies and the most comparable analytic approaches, the range of mean VSL estimates narrows considerably. All three meta-analyses report mean results from wage-risk studies in the United States, excluding studies that used the inappropriate Society of Actuaries data[7] and studies that covered only high-risk occupations (such as police work). The mean VSLs that the meta-analyses report for this group of U.S. wage-risk studies in the literature are in the range of $8-10 million.

Mrozek and Taylor argue, however, that many wage-risk studies have not included sufficient model specifications to account for unexplained differences in wages among industries and that this exclusion has caused upward bias in VSL results. They use their model to estimate the mean VSL if all the studies had used a more aggressive approach to control for differences in wages among industries. Their results suggest a mean VSL of about $3 million. Viscusi and Aldy point out that if unexplained differences in wages among industries correlate with differences in on-the-job fatalities, this could be part of the reason for the differences in wages. Thus, aggressive measures to include many industry dummy variables, for example, could cause downward bias in the risk coefficients in the wage function. The truth is probably somewhere in between—that is, the best mean VSL from the U.S. wage-risk studies is somewhere between $3 million and $10 million, but it is hard to say exactly where.

Kochi et al. provide important additional information on the VSL literature by including results from stated-preference studies. When considered alone, the stated-preference results reported by Kochi et al. have a mean VSL of about $3.3 million. That is at the lower end of the range of mean results of the wage-risk studies. Many of the stated-preference studies include subjects who are over 65 y old and thus reflect the preferences of a wider segment of the general population than the wage-risk studies, which are limited to working-age adults. The stated-preference studies also cover different types of mortality risks, primarily those related to traffic accidents and illness.

Another important difference is that the stated-preference studies examined by Kochi et al. include both U.S. studies and studies in other developed

---

[7]Society of Actuaries mortality data include deaths from all causes, so they are inappropriate for use in wage-risk studies; what is needed is a measure of the risk of on-the-job deaths.

countries. They do not report U.S.-only results of the stated-preference studies, as they do results of the risk-wage studies, and their overall mean reflects studies in all included countries. Although those are all countries with relatively high income (such as Canada and England), differences in per capita income, wealth, and other factors could result in systematically different WTP results. For a better comparison with the U.S. wage-risk studies, it is preferable to use results of U.S. stated-preference studies.

The authors provided a list of all the studies included in their analysis (R. Kramer, personal communication, 2006), which allowed further examination of this issue. The information they provided showed that only three of the eight stated-preference studies were done in the United States. Taking the results of those three studies (Viscusi et al. 1991; Hammitt and Graham 1999; Corso et al. 2001) and adding results of two additional studies done in the United States (Ludwig and Cook 2001; Alberini et al. 2004) allowed the calculation of a mean VSL for U.S. stated-preference studies. Alberini et al. report the results of a U.S. application of the same survey instrument used in the Canadian study reported by Krupnick et al. (2002), which was included in the Kochi et al. analysis. The Ludwig and Cook (2001) results were added because, although Kochi et al. listed the study as not providing a standard error of their mean results, Ludwig and Cook did report a confidence interval from which we were able to calculate an approximate standard error.

Some of the studies reported more than one VSL estimate based on different WTP questions or different model specifications. In those cases, an average result was calculated for each of the five studies. The weighted mean of the five U.S. estimates was then calculated by using the inverse of the standard error as the weight. The result was $6.2 million in 2006 dollars with a standard error of $1.2 million. That is considerably higher than the mean reported by Kochi et al. for all eight stated-preference studies and is close to the midpoint between Viscusi and Aldy's mean VSL from U.S. wage-risk studies and the adjusted mean VSL from U.S. wage-risk studies reported by Mrozek and Taylor (2002).

Kochi et al. report an overall mean VSL of about $5.4 million in 2000 dollars based on combining results from both types of studies. However, there are important questions about the usefulness of this combined mean VSL. The mean of both types of studies is to some extent an artifact of the number of wage-risk studies and the number of stated-preference studies. We see no reason for wage-risk results to be given more weight only because more of them have been done. If they measure something different, it is more important to ask which type of study is more likely to provide the VSL estimate that we want. It is an open empirical question as to why the results of wage-risk and stated-preference studies are different and whether the differences are statistically meaningful. It is not clear whether the differences in subjects' ages or in the causes of risk are what cause the apparent differences in results, or whether differences in methods explain the different results.

Empirical Evidence on Effect of Age on WTP to Reduce Mortality Risk

One of the most important questions related to estimating the benefits of reducing ozone-related mortality is how WTP to reduce mortality risk may vary with the age of the person at risk. Many of the available VSL estimates are for working-age adults, but illness-related mortality risk associated with ozone falls to a large extent on those over 65 y old. That is true even if the relative risks posed by ozone are the same for all ages, because the baseline risk of death from respiratory or cardiovascular illness is higher for those over 65 y old. This section summarizes the empirical evidence on how WTP for mortality-risk reduction varies with age, covering both stated and revealed preference studies.

Stated-Preference Results Bearing on Effect of Age on WTP to Reduce Mortality Risk

Krupnick (2007) has reviewed the stated-preference literature on the relationship between age and WTP for mortality-risk reduction. The review covers 36 studies of populations in the United States and abroad and includes such studies as those of Alberini et al. (2004) and Chestnut et al. (2004a) that examined populations in both Canada and the United States. Most of the studies examined values for an immediate risk reduction and in a private-goods context. Most covered representative samples of people of all ages or of people over some age (such as 40 y) or oversample people over 60 y old.

The age effects in this literature are measured according to a reference point: the difference (or percentage difference) in WTP for a given risk reduction in one age group compared with another age group. For example, the WTP of a 40 y old can be compared with that of a 70 y old, or the WTP of a group less than 65 y old can be compared with that of a group 65 y old and older. The effects are estimated either by studying the regression coefficient for WTP against age of the respondent after adjustment for other variables or by a subsample approach in which WTP is estimated separately for groups under 65 y old and over 65 y old and compared. The former approach, if correctly specified, estimates the effect of age independently of other factors, such as income, that may vary with age and also affect WTP. Stratification embeds all the factors that might correlate with age and also affect preferences for reducing mortality risks. The regression approach permits the analyst to test various specifications about how age affects WTP, including a quadratic specification that could test whether the expected inverted U-shaped relationship between WTP and age predicted in some theoretical analyses that used the life-cycle consumption model actually holds. The stratification approach often does not reveal those underlying relationships as cleanly, but it may not be important to know them. For instance, it may not matter in EPA's regulatory decisions whether the elderly value risk reduction less than younger people because of differences in income, wealth, age, or anything else. If age is the only factor being used to adjust VSL estimates for a population at risk, stratification by age gives sufficient information for making such adjustments. On the other hand, it may not be appropriate to make such adjustments if income, gender, or race effects were embedded in them.

The VSLs estimated by those studies range from $150,000 to $12 million (in 2006 purchasing-power parity-adjusted U.S. dollars), with an average of $2.7 million and a median of $1.7 million. (These figures exclude one study that had an extreme VSL of $58 million.)[8]

Of the 26 most reliable studies that tested for age effects, 14 reported evidence of a senior discount effect and 12 found either no effect or a senior premium. In those finding a discount, the size of the effect is difficult to present consistently because studies describe the reference group and the age of the seniors differently. However, these studies show a clustering in the 20-35% discount range for WTP at age 70 y vs WTP at age 40 y. DeShazo and Cameron (2004) found a significant discount in the same range in a comparison of 65-y-olds with 55-y-olds. However, WTP peaks at 55, rather than younger, as in Johannesson et al. (1997); and in other studies, the WTP is flat across the younger ages. Hammitt and Liu (2004) and Chestnut et al. (2004b) found a larger discount than the 20-35% cluster. Guria et al. (1999), with definite outlier results, found a far steeper discount. In absolute terms, the differences in the discount are wide. For instance, the gap between two studies is only 5%: a 35% discount (Johannesson et al. 1997) vs a 30% discount (Krupnick et al. 2007) at age 70 y relative to age 40 y; but in monetary terms, the senior discount is only about $400,000 for Krupnick et al. (2007) and almost $2 million for Johannesson et al. (1997). The fact that two teams of researchers—Chestnut et al. (2004a,b) and Alberini et al. (2004)—both surveyed large samples in both the United States and Canada and found significant senior-discount effects in only one of the countries (and for the opposite countries) does not instill confidence in the existence of a robust senior discount. However, both studies found negative (although insignificant) effects of age on WTP. If the null hypothesis were defined as the existence of a negative age effect, this hypothesis would not be rejected by the results of these two studies.

The review also reported results of a series of models to examine factors affecting the likelihood that a particular study would find a senior discount. In the entire sample, studies with a large sample are more likely to find a senior discount, and in alternative specifications, this variable is quite robust. The average age of the sample is also significant (at the 10% level) but less robust. Both results suggest that a senior discount can be found if the sample is large enough and contains a large enough number of seniors. Finally, a dummy variable indi-

[8]Comparisons of VSLs among studies (or comparisons of absolute rather than relative senior discounts) require conversion of VSLs to common units. Krupnick first converted non-U.S.-dollar-denominated study results into U.S. dollars by using study-year factors for purchasing-power parity and then scaled the VSLs to a common base year (2006, second quarter).

cating that a quadratic specification was used is nearly significant. That means that the quadratic specification, which allows a nonlinear relationship between WTP and age, is better than a linear specification at capturing any senior-discount effect that may exist. Overall, these results suggest that a deeper search for age effects through examination of alternative specifications for age might have found such effects.

These inconclusive findings are similar to the conclusions reached by Dockins et al. (2004), who, in summarizing other reviews performed by EPA, wrote that "empirical work exploring the relationship between age and the VSL has provided mixed results." They discussed the series of studies by Alberini et al. (2004) in the United States and Canada but otherwise examined only wage-risk studies. Another review of the literature (Evans and Smith 2006) offered the same conclusion, citing Alberini et al. (2004) and DeShazo and Cameron (2004): "These empirical analyses do not offer an unambiguous message."

Revealed-Preference Results Bearing on Effect of Age on WTP to Reduce Mortality Risk

Aldy and Viscusi (2007) reviewed the empirical wage-risk literature about the relationship between age and WTP to reduce mortality risk. They note that most wage-risk studies reveal little about the relationship between age and WTP for mortality-risk reduction even when an age-interaction variable is included in the specification. That is because the simpler specification implies a linear relationship with age, which is expected only under the most restrictive assumptions. Thus, an informative analysis requires at least a nonlinear specification that allows VSL to be either rising or falling with the worker's age. In addition, the authors note that rates of on-the-job fatalities also vary with age, so a comprehensive specification should take this into account. Although the overall on-the-job injury rate declines with age, the on-the-job mortality rate increases with age. Thus, both workers' preferences for risk reduction and employers' wage offers may vary with respect to workers' age; thus, there may be different wage-risk market equilibria for different age groups.

The authors report the results of their own analyses on this issue (Viscusi and Aldy 2007; Aldy and Viscusi in press) that allow the wage-risk equilibria to vary by age group. They find a consistent pattern of an inverted U-shaped relationship between VSL and age, with values for the youngest (18-24 y old) and oldest (55-62 y old) being roughly similar and the peak being about twice as high, usually in the middle age group (35-44 y old). Aldy and Viscusi (in press) note that the usual cross-sectional analysis does not take into account that expectations about future income and life expectancies vary with year of birth, both being higher in younger workers in the sample (given trends to higher income and longer life expectancies). They adjusted for those differences, and used the age-specific estimates of on-the-job fatality risk as described by Viscusi and Aldy (2007). They report that "the age-VSL relationship still follows an inverted-U shape, but accounting for year of birth effectively rotates the shape so that younger workers now have a lower VSL; the curve peaks a little later in life (at age 46), and older workers now have a higher VSL. The oldest worker in our sample (age 62) now has a VSL that is 35 percent lower than a 46-y-old worker's VSL, the peak value in this analysis."

Aldy and Viscusi (2007) note that if this result for the group of 55-62-y-olds is applied to all adults over 55 y old, they obtain a result similar to what EPA obtained when it estimated a lower VSL for those over 65 y old in a sensitivity analysis of the proposed Clear Skies legislation (which would create a mandatory program to reduce power plant emissions of sulfur dioxide, nitrogen oxides, and mercury by setting a national cap on each pollutant (EPA 2006b).

Aldy and Viscusi (2007) present several important conclusions from their review of the evidence in the wage-risk literature:

- VSL varies with age in an inverted U-shape, as has been predicted in some life-cycle consumption models.
- More flexible modeling that allows differences in both preferences and on-the-job risks according to age group results in estimates of VSL that decline less in older workers than was found in some of the earlier analyses.
- None of the wage-risk results support the assumption of a constant VSLY.

Implications of Results Regarding Age for VSLY

The weight of the evidence suggests that there is insufficient information to make a reasoned decision to stop using a single VSL and switch to either different VSLs for different ages, a constant VSLY, which imposes a linear (discounted) relationship between life years remaining and the VSL, or a VSLY that varies with age groups. Nailing down these effects will take large samples and thorough studies. What seems clear from the stated-preference literature is that in a private-goods context, there is no justification for use of a constant VSLY that is applied to all life years for all age groups. None of the study results are consistent with the assumptions that underlie a constant VSLY. These results do not rule out non-constant VSLs or VSLYs.

EPA's SAB (2007) reached similar conclusions and recommended against quantitative adjustments to VSL estimates by age because of inconsistencies in the available empirical results. However, it urged EPA to support more research on the question and argued that a variation in VSL with age should not be ruled out on theoretical grounds. It also recommended against a constant VSLY, noting that the empirical evidence does not support the assumption that VSL is proportional to remaining life expectancy, which is an implicit assumption when a constant VSLY is used.

We stress that the wage-risk studies discussed here do not include large numbers of people over 65 y of age, because many of them are not in the workforce. Thus, it is difficult to draw conclusions about how the VSL may differ in those over 65 y old from the wage-risk evidence because this would necessitate extrapolating beyond the age range in the studies.

Some analysts have advocated that EPA consider using quality-adjusted life years (QALYs) as a way of measuring mortality and morbidity risks in cost-benefit analyses. The QALY is a metric that blends life years saved with the quality of the life years when a person experiences acute or chronic disease. Death is generally given a value of 0, and perfect health a value of 1, although some indexes (for example, disability-adjusted life years [DALYs]) reverse this. Some QALY indices also permit scoring of health states that are "worse than death," that is, get less than a zero score. An Institute of Medicine (IOM 2006) panel commissioned by EPA with support from OMB to examine the role of QALYs in regulatory cost-benefit analyses made several recommendations that bear on the usefulness of this metric. The IOM panel noted that QALYs provide a useful perspective on the impacts of regulatory actions but that this is only one perspective. Other perspectives are also useful, such as monetary-benefit measures. The IOM panel was not asked to compare a life-years-saved metric with a lives-saved metric and did not do so. However, it did note, as has other literature (Hammitt, 2007), that cost-effectiveness with QALYs as the effectiveness measure should not be termed a "cost-utility" analysis, as in common practice, because to do so implies that the QALYs are a welfare measure (meaning, in a practical sense, that it would rank alternative projects in the same order as a benefits analysis). In fact, the conditions for this to be true are quite restrictive: for instance, life years need to be additive in the utility metric no matter who experiences them, how many are experienced by a given person, and when.

The IOM report recommended strongly against a common practice used to "convert" a QALY metric (life years lost) to dollars, specifically against putting a monetary value on a QALY. The concern was about mixing the medical-practice and welfare paradigms and about how the literature on monetary values does not support treating all life years saved as equally valuable (that is, a constant VSLY approach). A consequence of this medical paradigm, as voiced by the committee that authored the IOM report, is that QALYs implicitly put less value on mortality reductions in older people (because their life expectancy is lower than that of younger people's) and in those with existing chronic illness (because, owing to the quality adjustment, a life year extended for them is worth less than it would be if they were healthy). That practice is in direct conflict with the literature on WTP for mortality-risk reductions inasmuch as WTP is found to change less than proportionally with life expectancy and to be no different or even greater in the chronically ill compared with the healthy (see below).

OMB requires cost-effectiveness analysis in addition to cost-benefit analysis in regulatory-impact analyses of major rules and suggests that cost-effectiveness analysis be considered in all cases for comparisons of alternative regulations under discussion. Cost-effectiveness analysis defines the primary goals of the regulatory effort in some unit of measure that is not monetized. Costs of the alternatives are then compared on the basis of the cost per unit of benefit expected. For mortality risk, OMB notes that the units may be lives saved, life years saved, or QALYs, but it does not recommend one metric over the others. Unlike cost-benefit analysis, cost-effectiveness analysis does not allow calculation of net benefits, but it can help in selecting the most efficient regulatory alternative for achieving a given goal. Thus, it is most useful in comparing programs with goals that can be quantified with the same metric. When there are multiple important benefits of a regulatory effort and those other benefits can be monetized, they can be netted out from costs first, for a *net* cost-effectiveness analysis. Hubbell (2006) illustrates an attempt at cost-effectiveness analysis using QALYs for an air-pollution control rule (the heavy-duty engine and diesel-fuel rule). He included monetary valuation with a non-constant value per life year as an illustration but noted that there is a great deal of controversy regarding the appropriate way to infer welfare-based monetary values for VSLYs from the available VSL literature.

Empirical Evidence on Effect of Health Status on WTP to Reduce Mortality Risk

Few empirical studies have examined the effect of health status on WTP to reduce mortality risk. Alberini et al. (2004) and Krupnick et al. (2002) provide the most thorough examination of the question in stated-preference studies. They used the same survey instrument in the United States and in Canada and looked at the potential effect of several health-status measures on WTP responses, including family history of noncancer chronic disease, family history of cancer, personal history of hospitalization for heart or lung illness, personal history of high blood pressure, and widely used indices of physical and mental health (drawn from the SF-36 questionnaire). At times, a history of illness, either personally or in the family, was associated with higher WTP to reduce mortality risk, although not all the health measures were statistically significant and the significant measures differed between the U.S. and Canadian studies.

Underlying the question of the effect of health status on WTP to reduce mortality risk is the concern that people who are chronically very ill may have such a reduced quality of life that they do not attach as much value to a policy that would extend life. The studies noted above do not address that question well, because the samples are drawn from the general population and include few very ill people. One small study that sheds a little light on the question is by Tsevat et al. (1998). They asked hospitalized patients over 80 y old and their caregivers and family members to estimate tradeoffs that they would be willing to make between improving quality of life and extending life. The authors were surprised to find that patients put a higher value on extending life (relative to improving quality of life) than did the caregivers and family members. That suggests that the WTP to reduce mortality risk is not necessarily decreased and may be increased when quality of life is diminished.

Evidence on Effect of Risk Latency on WTP

There is little evidence on the effect of risk latency on WTP to reduce mortality risk, because few empirical studies have examined it. As illustrative of this literature, Alberini et al. (2006a) used results of their contingent-valuation studies in Canada and the United States to investigate the issue. The latent-risk question, which asked respondents under 60 y of age whether they were willing to pay some stated amount for a given reduction in risk starting at the age of 70 y and lasting 10 y, followed two questions on WTP for immediate risk reductions. They found that delaying the time at which the risk reduction would occur by 10-30 y reduced WTP by more than 60% in respondents in both samples of people 40-60 y old. The implicit discount rates are equal to 3.0-8.6% in Canada and 1.3-5.6% in the United States, in contrast with earlier estimates of the discount rate in risk-reduction tradeoffs, which ranged from 0.3% (Johannesson and Johansson 1996) to 14% (Viscusi and Moore 1989).

Studies to Estimate WTP for Increases in Life Expectancy

Analysts have pointed out that quantifying the effect of changes in pollution exposure as annual changes in numbers of deaths in a population obscures the dynamic dimension of the change in exposure and its overall effect on mortality over time (e.g., Brunekreef et al. 2007). A more comprehensive way to describe the effect is as a shift in the survival curve over time. For the individual, that means an increase in the probability of surviving over all future periods and thus extending life expectancy. Of course, death is not prevented for the individual; rather, future survival probabilities increase, and death is more likely to occur at a later age. This can be summed up as a change in life expectancy, but it may also be described as a change in annual mortality rates over some extended period. Chapter 4 notes some unresolved issues about how best to characterize the change in mortality risk in the population as a result of a reduction in pollution exposure. In particular, several authors have raised concerns that the number of deaths prevented is not a stable measure over time (e.g., Miller and Hurley 2003; Rabl 2006) and have argued that a better measure would be the increase in life expectancy or life years saved. Economic valuation of the reduction in mortality rate can be derived directly from available WTP studies, which provide WTP values to reduce annual mortality risks. However, economic valuation of changes in life expectancy are not as straightforward and, with stated-preference approaches, are difficult to communicate to respondents, although these issues could perhaps be addressed with further empirical research.

A few stated-preference studies have tried to estimate WTP values for changes in mortality risk that are presented to respondents as changes in life expectancy (Johannesson and Johansson 1996; Johnson et al. 1998; Morris and Hammitt 2001). To define a change in life expectancy that is the same for everyone, those studies have defined a change in risk that starts at a future age, such as 60 or 70 y, and asked respondents what they would pay now for this future risk reduction that increases their life expectancy by some specified amount. The results are difficult to interpret. Comments from respondents suggest that some share of them dismiss the risk reductions as too small to be worthwhile. There appears to be a tendency to view an increase in life expectancy as adding time at the end of life when quality of life may be diminished rather than recognizing that an increase in life expectancy means an increase in the probability of survival in all future periods. The WTP questions seem to be additionally confounded with the presentation of a risk reduction that does not begin to have benefits until some years in the future. Morris and Hammitt (2001) made an interesting comparison between presenting a continuing annual risk reduction (to half the respondents) and an equivalent increase in life expectancy (to the other half of the respondents), both starting at the age of 60 or 70 y for a hypothetical pneumonia vaccine. They presented a 0.2% annual risk reduction starting at the age of 60 y or an 11-month increase in life expectancy (after getting the vaccine at the age of 60 y); for another portion of the sample, they presented a 0.2% annual risk reduction starting at the age of 70 y or a 5-month increase in life expectancy (after getting the vaccine at the age of 70 y). The substantial lag between the payment and the beginning of the risk reduction (an average of 25 y in the sample, which had a mean age of 40 y) confounds the interpretation of the results. The WTP results for the two risk presentations were not very different in those who said they would consider paying now for the future vaccine, but respondents who received the life-expectancy version were more likely to say that they would not consider getting the vaccine (33% vs 26%). The most common explanation was that the benefit was too small. Compared with many risk-reduction efforts that people routinely undertake (such as cancer screening and wearing seat belts), the benefit is actually large, so the response raises some questions about the presentation. For example, Corso et al. (2001) demonstrated that visual aids are useful and needed to help respondents to understand quantitative risk information.

Alberini et al. (2006b) report VSLY estimates derived from WTP values obtained in a stated-preference study conducted in France, Italy, and the UK regarding annual mortality-risk changes over a 10-y period. The VSLs for the 5-in-1,000 risk change over 10 y were a mean of 2.3 million euros and a median of 1.1 million euros. Using subjects' own perceptions of their life expectancies, the authors calculated the implied VSLY and obtained a mean of 142,000 euros and a median of 58,200 euros. However, they found that the VSLY was significantly higher in those 60 y old and older (27% above those 40-49 y old). It is notable that the 10-y mortality-risk change converts to a different life-expectancy change in each age-sex cohort. It remains unclear whether the conversion from a WTP for the risk change to a VSLY for the associated change in life expectancy is the appropriate way to obtain this value.

Analysts in Europe (Bickel and Friedrich 2005) report using results from Alberini et al. (2006b) as the basis of their estimates of VSLY in the 2005 update of the method for estimating externalities of energy. They used the estimates after presenting their argument that it is preferable to estimate the mortality effect of air pollutants on the basis of changes in life expectancy. Their selected central estimate of 50,000 euros per life year was derived from the median WTP value for the 5-in-1,000 mortality-risk change over 10 y (median VSL, about 1.1 million euros). The ExternE authors converted the 10-y mortality-risk change to its equivalent in increased life expectancy for each age-sex cohort and calculated the VSLY implicit in the WTP responses. It is notable that their selected central estimate was based on one of the lower WTP results because it was based on the median rather than the mean value and on the 5-in-1,000 10-y risk reduction rather than the 1-in-1,000 10-y risk reduction. The VSLY based on the mean value for the 5-in-1,000 reduction was about 125,250 euros (the associated VSL was about 2.3 million euros). The authors note that the WTP questions in this study include an implicit latency in that the risk reduction occurs over a 10-y period. Using an assumed 3% discount rate, they calculated that the undiscounted VSLY based on the median VSL would be about 75,000 euros rather than 50,000 euros. They concluded that the latency was appropriate for valuation of a reduction in mortality risk posed by long-term PM exposure. As a high estimate, they used 150,000 euros, derived from the WTP responses in the UK survey to the 1-in-1,000 10 y risk reduction (the associated VSL was about 3.3 million euros).

They based their low estimate on results from the study in France, which included a WTP question specifically for changes in life expectancy. Details of the estimation approach have not been published, so it is difficult to evaluate. Desaigues et al. (2004) reported VSLY estimates ranging from 20,000 to 220,000 euros, depending on the underlying change in life expectancy implied in the different questions included in the French survey. The ExternE analysts note that the WTP question for life-expectancy increases stressed that it did not mean just a few months at the end of life: "A crucial point that needs to be explained very carefully in such a questionnaire is that air pollution mortality does not cut off a few months of misery at the end of life but causes 'accelerated aging'" (Bickel and Friedrich 2005, page 44). The low estimate of VSLY selected by the authors for use in the recent ExternE assessment is a value of about 20,000 euros per year of additional life expectancy. Although the ExternE analysts selected a range of VSLY estimates, they used an approach that presumes a constant VSLY, which ignores the evidence that this value is likely to vary with age.

FINDINGS AND RECOMMENDATIONS

Finding: The charge to this committee concerns monetary valuation of mortality-risk reduction for regulatory impact analyses that are based on the basic premises of cost-benefit analysis. In this context, therefore, we focus on WTP values for mortality-risk reduction. Cost-benefit analysis, however, focuses on economic efficiency, and many other ethical and legal factors are appropriate to consider in policy and regulatory decisions. In general, these issues should be considered separate from the cost-benefit analysis rather than interjected into the valuation estimates because this endangers the neutrality of the analysis. However, there may be some instances when such interjection is appropriate. For example, the decision by EPA to not adjust WTP estimates for local differences in income levels is justified because to do so creates a situation that favors greater environmental protection in wealthier locations, an outcome policy makers judge to be unfair and in many cases illegal.

The committee stresses that government decision-makers need information on how the WTP for mortality-risk reductions varies by risk characteristics, population characteristics, and for both risk as a private good and as a public good. How they choose to use this information, however, is not strictly a technical decision, but depends on ethical precepts, legal precedent, the quality of the evidence and other factors that may be beyond the analysts' control or purview.

Recommendation: We recommend that this finding of the committee be considered within OMB and the agencies that use monetary values for mortality-risk reductions in their regulatory analyses. These agencies should develop a plan for generating the information needed to determine how WTP varies for different populations and different risk contexts. In addition, there should be an exploration and debate to determine the appropriate uses of this information. Such a debate should go beyond economic considerations and include ethical and public policy perspectives.

Finding: Both economic theory and the empirical evidence are inconclusive about how individuals' WTP for reducing their own risk vary with two important individual characteristics: age as a proxy for remaining life expectancy and health status. Although we conclude that the empirical evidence is insufficient to support a specific quantitative adjustment to the WTP for reduction in annual mortality risk based on differences in remaining life expectancy, we do not reject the concept that such adjustment may be appropriate. It is plausible that people with less remaining life expectancy are willing to devote less of their resources to reducing their mortality risk than those with more remaining life expectancy.

Characteristics of the risk that may affect individual WTP values include the type of risk (such as illness or accident) and its latency. The literature is inconclusive about the influence of risk characteristics (see below). The effects of latency on WTP values are straightforward conceptually (the WTP for a future death-risk reduction should be less than the WTP for an equivalent immediate death-risk reduction), and there is some published support for such estimates. However, the epidemiology literature is not sufficient to estimate the degree of latency in mortality response to ozone exposure.

Recommendation: Despite many concerns about the accuracy of using the same WTP value (or range of values) for all mortality-risk reductions, we recommend this, with appropriate scaling to the size of the risk change, as the most scientifically supportable approach for monetary valuation of ozone-related mortality, given the currently available information in the economics and epidemiology literatures. Empirical evidence of how WTP varies with population or risk characteristics is not sufficiently consistent to support a change in this practice that EPA has been using for many years. Researchers should continue to explore how WTP for mortality-risk reduction may vary with personal characteristics (such as age and health status), with the type of risk (such as accident and illness), and with its latency. This most likely implies more stated-preference studies to include the elderly population.

Finding: The use of a constant value for life years saved in the valuation of increases in life expectancy requires the assumption that WTP values for mortality risk reductions be consistently declining with increasing age. The empirical evidence does not support that assumption and therefore does not support the use of a constant VSLY. The literature does not reject the use of a non-constant VSLY, however. And the epidemiological literature (see Chapter 4) may favor reporting life years saved.

Recommendation: Unless future research produces empirical support for the assumptions that underlie a constant VSLY, EPA should not attempt to adjust for remaining life expectancy by calculating life years saved and applying a constant VSLY. It may be appropriate to calculate and report life years saved (in addition to reporting lives saved or changes in annual mortality rates), but it is not appropriate to use a constant VSLY as a monetary valuation of life years saved, except perhaps in a bounding exercise. The committee cautions against use of such an analysis in anything but a sensitivity analysis, however (see below). There is likely to be good reason to use a non-constant VSLY or a non-constant VSL, once the empirical literature is sufficient to support this transition. The committee stresses, however, that the status quo of using a uniform VSL should be continued until there is sufficient empirical evidence of how WTP for mortality-risk reduction varies with differences in remaining life expectancy and other factors, which the committee concludes is not yet available.

Finding: Most of the revealed-preference and stated-preference studies relied on by EPA to obtain estimates of the VSL have estimated these values for a context (such as traffic accidents or workplace accidents) and for a population that differs, for example, by age, health status, and income, from the population facing the pollution-related risks that EPA is assessing. Applying the available estimates in EPA's assessments in different contexts of risks (such as illness vs accident) and population characteristics (such as age) introduces considerable uncertainty about how these factors affect the average WTP values. However, the current literature is inconclusive as to how and how much the WTP values may vary with those factors.

Recommendation: The use of average WTP estimates selected from the literature should reflect results from both revealed-preference and stated-preference studies, take into account the strengths and weaknesses of each approach, and consider how closely the studies match the policy context in population at risk and type of risk. Given the limited studies available for different risk contexts, it is difficult to say how much the WTP values may differ, but wage-risk studies are a poor match to the population and the risk context for the ozone-mortality case. EPA should give less weight to these studies in selecting WTP estimates than it has in the past.

Finding: The direction of the expected error, if any, in using the average VSL in the literature for valuing changes in ozone-related mortality risk is more likely to be toward overstating the WTP to reduce this risk. That is because greater mortality risk associated with ozone appears to be in the elderly population and because latent risk may be involved. The remaining life expectancy in this population is substantially less than that in the population as a whole, and its WTP to reduce mortality risk might also be less. However, the lower WTP as a result of lower remaining life expectancy in the elderly may be offset to some extent by higher WTP because of poorer health status or higher baseline risk. Although results in the empirical literature are not consistent, several studies suggest that WTP to reduce mortality risk does not change or declines slightly with age. The evidence suggests that, for ozone, a proportional adjustment of the VSL for remaining life expectancy (that is, using a constant VSLY) would result in WTP that is too low.

Recommendation: Given the uncertainty in available VSL estimates for ozone-related mortality, it is appropriate to conduct some sensitivity analyses with alternative estimates or assumptions. The purpose of sensitivity analyses is to see whether the overall conclusions of the cost-benefit comparison are changed, for example, whether net benefits are still positive under alternative economic-valuation assumptions. In general, there is less confidence in the sensitivity analyses because the alternative assumptions are more speculative than the primary assumptions or deviate from the status quo, which the committee feels puts a burden of proof on those who would overturn it. By definition, sensitivity analyses should be given less weight in the presentation of results. However, they can be included in the summary and conclusions. The selection of alternative assumptions for the sensitivity analyses can be based on either theory or evidence. For example, different published empirical estimates of the relationship between WTP for mortality-risk reduction and age could be selected as illustrative of the range of published results, including estimates of VSLY or VSL that vary with age.

RECOMMENDATIONS FOR FUTURE RESEARCH

The above recommendations for EPA imply research needs for estimating the WTP for changes in mortality risk and for changes in life expectancy. They also imply that the research primarily should use stated-preference methods, although researchers may find ways to address the issues with further development of the revealed-preference approach.

A fundamental need is to understand better how age and remaining life expectancy affect WTP for reductions in mortality risks or increases in life expectancy. One step would be to ask that future (or even previous) studies report total age effects—that is, WTP by age cohort—in addition to marginal effects of age on WTP. Given the correlation of age with some of the other factors, there may be less uncertainty in the estimates of a total age effect. However, age-related income differences, sex differences, health differences, and the like would then be embedded in the estimates, and it might not be appropriate to use different VSLs that have these effects embedded.

Several recent studies (e.g., Alberini et al. 2004) have looked qualitatively or quantitatively across the valuation literature to assess age and other effects. The assessments have been hampered by issues in the reporting of information and the lack of availability of the datasets produced. Future research funded by EPA should urge that its datasets be made available for meta-analysis.

More fundamental research is needed to explore and develop methods for communicating and valuing changes in mortality risk that reflect the full life cycle. Studies to date have focused on WTP for annual changes in mortality risk, but the risk change of interest in most pollution control assessments is more comprehensively described as a shift in the survival curves, which are plots of survival probability in all future periods and from which life expectancy is derived. That poses a challenge for stated-preference studies. A few studies have used changes in life expectancy to define mortality-risk change (e.g., Johannesson and Johansson 1996; Morris and Hammitt 2001), but many respondents seem to dismiss these changes as time added at the end of life, when quality of life may be substantially reduced. DeShazo and Cameron (2004) included multiple periods in their presentation of risk changes in an ambitious and innovative stated-preference study; analysis of their results has not yet been published.

It is also important to learn more about how mortality-risk characteristics affect the valuation of reducing risk. Environmental-benefits assessments have relied primarily on estimates of WTP to reduce risk of accidental death to estimate values for reducing risk of illness-related death. A few studies have compared values of WTP (or nonmonetary preferences) to reduce risk in different contexts (e.g., Magat et al. 1996; DeShazo and Cameron 2004), and more studies along these lines are needed.

Studies in a public-goods context that can isolate the effect of the age of those who benefited on the WTP of respondents for programs that reduce mortality risk in the community might help to resolve this issue as well. But in addition to such studies, we would need approaches to distinguish paternalistic from non-paternalistic altruism (as defined in Jones-Lee 1991) if double-counting of benefits is to be avoided. Conceptual analysis would be needed to understand how such results might be appropriately used in cost-benefit analysis, in which the usual paradigm is to sum private WTPs of the beneficiaries.

6

Overall Conclusions and Recommendations

The committee's statement of task (see Chapter 1) presents specific questions that were addressed by the committee in Chapters 3-5. This chapter discusses the overall conclusions and recommendations developed in the earlier chapters in the context of the regulatory benefits-assessment process. It also discusses implications for future Environmental Protection Agency (EPA) regulatory impact analyses (RIAs) that include changes in ambient ozone concentration.[1]

Part of the committee's charge was to evaluate recent analyses of epidemiologic studies that found a modest but consistent relationship between short-term ozone exposure and premature mortality. During its deliberations, the committee was mindful of the information needs and framework for benefits assessment. There is a fundamental difference between information needed for regulatory benefits assessment and information needed to set a protective health standard. Selection of a health-based standard focuses on the lowest ambient concentration that poses a risk of adverse health effects in the most sensitive population, whereas benefits assessment uses information to estimate all the health-risk reductions in the entire population that is expected to experience a change in ambient concentrations.

OZONE MORTALITY EFFECT

In carrying out its charge, the committee considered not only the recent epidemiologic evidence, but also toxicologic and pathophysiologic evidence that points to mechanisms by which ozone may contribute to premature mortality. As a gas, ozone is highly reactive, and once inhaled it is immediately engaged in the respiratory tract by the epithelial fluids and cellular membranes that it contacts.

---

[1]In this report, *ozone* is used to refer to the broad array of photochemical oxidants in ambient air, of which ozone is the primary component.

Because ozone does not penetrate cells but leads to pulmonary and nonpulmonary events, a cascade mechanism has often been proposed to account for its toxicity. Human chamber and toxicologic studies have yielded strong evidence indicating that short-term exposure to ozone can exacerbate lung conditions, causing illness and hospitalization, and potentially lead to death. The available evidence on ozone exposure and exacerbation of heart conditions, which is less abundant, points to another concern.

Epidemiologic studies also have found that exposure to ozone (as an indicator of the broader mix of photochemical oxidants) is associated with those effects. Although methodologically somewhat different, the studies have been consistent in their use of large datasets with consistent diagnostic codes for health end points, nationally available ambient air measures, and data on adjustment for some potential confounders. The committee found that the four recent time-series analyses and meta-analyses of the relationship between exposure to ozone and premature mortality add to that evidence by providing robust statistical evidence of an association (Bell et al. 2004, 2005; Ito et al. 2005; Levy et al. 2005).

On the basis of the additional insights obtained from its review of the new time-series studies and its review of the broader evidence, the committee concludes that short-term exposure to ambient ozone is likely to contribute to premature deaths. Despite some continuing questions about the evidence, the committee concludes that it is strong enough to be used in the estimation of the expected mortality risk reduction that would result from reduction in exposure to ozone or the photochemical-oxidant mixture.

In its RIA for the finalized ozone national ambient air quality standards (NAAQS), EPA (2008b) analyzed a variety of assumptions about the association between ozone exposure and premature mortality. They included the assumption that the association is not causal (see Appendix B). Although it is rarely possible to exclude the possibility of zero effect in such analyses, the committee concludes that an absence of any effect is unlikely.

INTERPRETATION OF RESULTS OF HEALTH STUDIES

Those who evaluate regulatory benefits seek information from health researchers, for example,

- To what extent is the relationship between ambient ozone concentration and mortality response due to ozone as opposed to copollutants that are quantified separately, such as airborne particulate matter (PM)?
- Given that age and health status are important in estimating the quantity and quality of the remaining life expectancy, how dependent is the ozone-mortality relationship on those and other personal characteristics, such as socioeconomic status?

Studies relating measurements of ambient ozone to health end points must deal with formidable obstacles because the dynamics of ozone and associated pollutants are complex, particularly the effects of temporal and spatial variations on individual human exposure. The committee reviews below the major factors that it considered in its review of the evidence. These factors can affect estimates of risk of ozone mortality in various ways. In some cases, the factors would cause an underestimation of risk; in other cases, an overestimation. On balance, the committee considers the evidence from the studies to be strong enough for making risk estimates, but the various factors and their potential effects on the estimates should be fully acknowledged.

Estimating Exposure to Ambient Ozone

Time-series epidemiologic studies of ozone typically characterize exposures in terms of ambient concentrations at a centrally located monitoring site or at several sites in the study area. That use of ambient concentrations to indicate ozone exposures is a source of considerable uncertainty related to how well they reflect actual ozone exposures and how well ozone effects can be separated from effects of other pollutants or weather conditions.

The magnitude of exposure errors probably varies with several factors, including season, home ventilation characteristics, and exposure averaging time. Ozone's mortality effects, for example, were shown to be stronger in the warm season than in winter or the entire year.

Over 24 h, personal ozone exposures are weakly associated with corresponding ambient concentrations, and the association is stronger in summer than in winter. The seasonal variability probably reflects increased home ventilation in the hotter summer months. Even in well-ventilated conditions, however, the ozone association is such that only minor changes in indoor ozone exposures are expected to occur in response to moderate changes in outdoor concentrations. **Together, low personal exposures and weak relationships between personal exposure concentrations and ambient concentrations suggest that 24-h ambient ozone concentrations are poor proxies for personal exposures.**

For shorter averaging periods, such as the afternoon (when both personal outdoor activity and ozone concentration can be at their highest), results from a scripted exposure study suggest that hourly or daily peak ambient ozone concentration may be an appropriate proxy for corresponding hourly or peak personal exposure. Whether observations from this study are relevant for individuals at risk of ozone-related death warrants further examination. **Personal ozone exposure is a major source of uncertainty in ozone-mortality risk estimates; additional studies are needed to determine whether using the shorter averaging times reduces this uncertainty. These studies will require the development of new measurement methods that have sufficient sensitivity to measure these likely low, short-term exposures.** Additional studies assessing personal exposure-ambient concentration relationships may not only reduce uncertainties, but might help to explain variability in ozone-mortality findings across studies (that is, by determining how the strength of the correlation and the extent to which personal exposures change in response to changes in ambient concentrations varies across cities or subpopulations).

In a given location, various ozone metrics (averaging periods) tend to be highly correlated, so it may seem somewhat unimportant to distinguish between them, and the degree of correlation may make it difficult for an epidemiologic study to do so. However, the choice of metric can have a larger effect on estimates of expected benefits of control programs. For example, a program that lowers emissions of oxides of nitrogen (NO_x) could reduce peak ozone concentrations but raise average concentrations. In that case, a cost-benefit analysis based on an association between premature mortality and average ozone could appear to have a negative effect, whereas an analysis based on an association with peak summertime ozone could show a benefit. It is unknown which is more accurate, so evaluating the benefits of urban NO_x reductions is uncertain both in magnitude and in direction. The choice of exposure metric can also be important in designing a strategy to control ozone-precursor emissions (see Chapter 3).

Although the recent time-series studies have to the extent possible included analyses of alternative ozone metrics (such as 1-h maximum or 8-h maximum), it is important to examine the relationships further to examine dose-response relationships, test potential confounding more effectively, and inform future regulatory choices among different actions that might have different effects on peak and multihour averages.

Potential Confounding by Other Pollutants

The committee found that short-term ozone exposure is likely to contribute to premature mortality in addition to the risks posed by weather and PM, but studies to date have not been sufficient to control for potential confounding by or interactions with condensed-phase constituents of airborne PM, such as sulfates, acids, elemental carbon, and metals. **Colinearity among ambient pollutants raises concerns about possible confounding of ozone mortality effects by correlated copollutants.** The concerns center on the possibility that effects associated with ambient ozone may be the consequences not only of ozone exposure but of correlated pollutants not included in the health-effects model. Of the possible confounders, ambient $PM_{2.5}$ (PM with a diameter no greater than 2.5 µm) and weather have raised the most concerns about confounding.

Strong summer correlations between ozone and $PM_{2.5}$ in some locations may be attributed to similarities in their formation. However, associations between ambient ozone and $PM_{2.5}$ differ substantially in the western and eastern United States, between summer and winter, and for different PM components. Also, correlations among ambient ozone and other pollutants have been shown to vary by averaging period. These results suggest that the potential for confounding of ozone-health effects also varies by these factors, as discussed in Chapter 4.

The potential for confounding by ambient $PM_{2.5}$ probably depends also on particle composition, which varies by location and season. Studies have suggested that health effects of $PM_{2.5}$ may differ by component, so confounding might be an effect of specific $PM_{2.5}$ components, such as sulfate, elemental and organic carbon, metals, and secondary organics; this matter should be examined. It will be difficult to address such confounding with currently available data, because $PM_{2.5}$ component data have only recently been collected routinely at many sites and because winter time ozone is not often measured. In the near future, the Speciation Trends Network (STN) may provide data on new measurements by species. However, unlike data on ozone, such data are generally available only once every 3-6 d (see Chapter 3). More-frequent measurements may be needed to improve understanding of how short-term variation in $PM_{2.5}$ components might confound ozone-mortality associations.

Assessing Ozone-Mortality Relationships During Winter Months

There is a lack of observed association between ozone and mortality during periods when ozone is low, such as winter. Reasons for this lack of association are not well understood in part because of the decrease in monitoring during such periods. Better understanding of the association is important for a full exploration of (1) seasonal differences in risk, (2) how these seasonal risk differences vary spatially between communities with warmer and cooler winters, and (3) ozone-mortality relationships at lower ozone concentrations.

Ambient ozone is one of the most well-characterized pollutants in the United States, but ozone monitors in many locations are operated only during the ozone season, which varies from city to city. During that time, ozone monitors provide nearly continuous measurement, although concentrations are typically reported hourly.

Threshold

To characterize the association between daily variations in ambient ozone concentrations and daily variations in deaths, a linear model is used in which it is assumed that the change in mortality risk is constant across pollution concentrations. It is unlikely that the association between exposure and response at the individual level follows that simple mathematical formulation. Individuals have their own susceptibility, characterized by a unique exposure-response association. That association may be characterized by a particular threshold, a concentration of exposure to ozone below which there is no added risk of death. A person's threshold will vary, depending on the person's "frailty" at any given moment and thresholds may depend on the averaging period used to assess exposure.

The time-series design, however, relates individual exposure not to individual risk but to the average of ambient concentrations for the "at risk" population on any given day. Thus, the appropriate concentration-response function for the time-series studies is based on an aggregation of individual exposure-response curves. Aggregation of a large number of complex functions can yield smoother and more nearly linear curves at the population level, so we would expect the concentration-response function based on time-series studies to have a form that is relatively simple.

Those results suggest a near-linear association between ambient concentrations of ozone and daily mortality in the United States. Exposure misclassification caused by use of an average of ambient fixed-site monitoring data to estimate population-average personal exposure makes it more difficult to distinguish between linear and threshold models. Estimates of the concentration-response curve based on epidemiologic studies with imprecisely measured exposure should be viewed with caution. A sensitivity analysis of the shape of the curve may be required to capture the uncertainties in this procedure. Moreover, approaches based on 24-h averaging may cloud thresholds related to actual exposures which may be better represented with shorter averaging times, such as by using an 8-h maximum or 1-h maximum.

On the basis of its review of the evidence, the committee concludes that the association between short-term changes in ozone concentrations and mortality is generally linear throughout most of the concentration range, although uncertainties make it difficult to determine whether there is a threshold for the association at the lower end of the range. If there is a threshold, it is probably at a concentration below the current ambient air quality standard.

Susceptibility and the Interpretation of Mortality Studies

The evidence presented in Chapter 4 leads to the preliminary conclusion that the effects of ozone on acute death rates are likely to be larger among those with pre-existing disease and the list of plausible effect modifiers is rather long, although insufficiently investigated at this time. The role of genes in ozone mortality has not been investigated, but they undoubtedly play an important role as modifiers of various pathways involved in the effect. Interaction of the various factors has not been investigated.

One can infer that ozone mortality depends on subjects' susceptibility profile, which consists of a wide range and combination of factors. Susceptibility is most likely not a dichotomous trait but a dynamic characteristic that follows a wide distribution in the population, from no to high susceptibility. The level of susceptibility of any subject at any time may depend on the distribution of various exogenous and endogenous determinants of susceptibility (and their interactions) and their change over time.

The observable association between ozone and mortality reflects the weighted average of all true (but unobservable) individual concentration-response functions. The latter cannot be established, but the risk of dying because of a 10-ppb increase in ozone concentration is likely to be substantially larger in the most susceptible persons than in the general population.

Although susceptibility factors certainly matter, the distribution of the ozone-mortality effect estimates among groups with different categories of susceptibility is not known; that is, the quantitative details of the heterogeneity of effects are not readily available. Consequently, the overall (population-weighted average) effect in the total population is the only currently scientifically supportable approach for use in risk assessment. Its use is appropriate, even though the population-weighted mean effect may not be a valid estimate for any specific sub-population, because the actual effect may be much larger among the susceptible individuals but much smaller or zero among the less susceptible. However, that is a source of an unknown amount of uncertainty when one is calculating the benefits of a reduction in ozone. If the hazard and risk distributions[2] in the population are not independent, estimates of excess deaths and of life years lost in the entire population that are attributable to changes in ozone exposure based on time-series studies will not equal the averages of these quantities in all risk groups. The size of the difference is not known but could be estimated if risks for susceptible groups are obtained (see Chapter 4).

Short-Term Mortality Displacement

Is there evidence to say what share of the mortality associated with ozone in the daily time-series studies may be very short-term displacement of deaths that would have occurred within days or weeks in the absence of ozone exposure?[3] Current economic valuations of mortality are based on study samples that have average remaining life expectancies. It would be problematic to extrapolate these valuations to a circumstance in which the remaining life expectancy of the population most at risk is very short. When a person is aware that death is imminent and quality of life is seriously compromised (for example, the person is bed-ridden, is in great pain, or has extremely diminished cognitive function), extending life by a small amount of time may not have a high value. But if the person can be treated and the condition improved and an acceptable quality of

[2]In this context, the distribution of hazards among the population is the likelihood of death at any age; the distribution of risk is the association between ozone and death at any age.

[3]The term *harvesting* is sometimes used instead of *short-term mortality displacement* to refer to the concept that air pollution leads to the death of people who are highly susceptible and near death (and die a few days earlier than they would have without air-pollution exposure) rather than the death of people who are not otherwise near death.

life can be restored, extending life may have a high value even if the person has other health impairments or is quite elderly.

Given the design of acute-effects studies, the time lost or the prematurity of deaths is not directly estimable without possibly unrealistic assumptions. In the absence of knowledge about the amount of time lost because of acute effects, one may hypothesize that air pollutants were able to trigger death only in a pool of very frail people, those already in very bad health. Assuming that the remaining life expectancy of those frail people was very short even in the absence of pollution, the effect of air pollution would consist of only a minor shift of the time of death, namely, a short advancement of death.

On the basis of available evidence, the committee concludes that deaths related to exposure to ozone (and other photochemical oxidants) are not restricted to people who are at very high risk of death within a few days. The evidence comes from the recent analysis of time-series in several U.S. cities that focused on identifying a mortality-displacement pattern in the time course of exposure and death (Zanobetti and Schwartz 2008). In that analysis, it was clear that short-term mortality displacement could not fully explain the observed increase in death; short-term ozone exposure was likely to have contributed to shortening the lives of people who had compromised health, not necessarily just those near death. That evidence is based on results of only one study, however, and warrants confirmation by other studies.

Distributed Lag

Deaths related to short-term ozone exposure may not occur until several days after exposure or may be associated with multiple short-term exposures. Many studies of short-term effects investigate the change in death rates for only one or a few days, but distributed-lag models look further ahead to capture delayed acute effects, often referred to as subacute effects. Such analyses are useful in understanding the statistical distribution of time between an increased concentration of ambient ozone and the time pattern of occurrence of death. On a population level, it is extremely unlikely that all ozone-related deaths occur either immediately or within 1-2 d of exposure. A more plausible model assumes susceptibilities to death (or frailty) and the (competing) success of intervention strategies to follow a distribution in which some people die immediately or within 1-2 d whereas others first suffer acute ailments (such as a myocardial infarction or pneumonia) and death may be delayed by partially successful treatment or occur as a result of the decompensation of defense mechanisms.

In contrast with the PM health-effects literature, few data are available that are based on using distributed-lag models of ozone mortality. Those available suggest that effect estimates for both ozone and PM steadily increase with increasing time of the investigated effects. Specifically, subacute (longer-term) effects that are combinations of effects of several days or weeks of exposure are larger than immediate short-term effects, and estimates based on cohort studies are the largest. However, the increase in the estimates is far larger for PM than for ozone; this may be a consequence of attenuation due to large errors in characterizing exposure to ozone, or it may reflect a more dominant role of acute disease related to ozone exposure whereas the role of chronic disease is more important in the effects of PM. **As has been the case with PM, analyses conducted with distributed-lag models over several days appear to capture the overall effects of ozone better than same-day data, but there have been relatively few of them and further confirmation is warranted.**

Chronic Exposure

Long-term effects of ozone on mortality are considered to be the result of cumulative effects on people who have chronic disease caused by repeated exposure to ozone. Although some inconsistencies remain to be clarified, the observed associations between ozone exposure and decreased small-airway lung function during childhood and adolescence suggest that ozone-related mortality is at least partially attributable to exposures across a period of more than a few days. Ozone tissue dose is highest in the small airways, so the findings are in line with expected ozone-related conditions that result in reduced lung function. The evidence of an effect of long-term ozone exposure on lung-function growth increases the plausibility of its effect on mortality. The association between poor lung function and life expectancy is strong and well established.

The standard approach to investigating effects of cumulative ozone exposure on life expectancy is the cohort study, in which large numbers of subjects are followed for several years. After taking into account all other factors that are likely to affect mortality, cohort studies can test the null hypothesis that mortality is the same among populations that have different ozone-exposure histories. However, none of the cohort studies available at this time were designed to investigate chronic effects of ozone, and differences in ozone exposure among subjects in each study tended to be rather small.

Several North American and European air-pollution cohort studies have focused on ambient PM or markers of local traffic to characterize exposure of study cohorts. Assessments of the association between life expectancy and ambient ozone have been far less extensive. Analyses of the most extensive of the cohorts (that of the American Cancer Society) have found small positive effects in warm seasons, but other cohort studies have not found positive associations between long-term average ozone concentrations and cardiopulmonary mortality after controlling for $PM_{2.5}$.

The weak current evidence from cohort studies of an association of premature mortality with chronic exposure to ozone suggests that risks are larger than those observed in acute effects studies alone. If further confirmed, the evidence from cohort studies of an association of premature mortality with longer-term exposure would tend to support the notion that effects seen in time-series studies reflect only a portion of the total effect.

The use of large cohorts with long followup periods may be required because the long-term ozone-exposure mortality risk appears to be much smaller than that associated with $PM_{2.5}$ It may be necessary to pool information from several existing cohorts to obtain sufficient statistical power.

HEALTH-BASED INFORMATION FOR BENEFITS ASSESSMENTS

Mortality Time-Series Results

The committee recommends that ozone-related mortality be included in future estimates of health benefits of ozone reduction. The committee further recommends that the greatest emphasis be placed on estimates based on systematic new multicity analyses using national databases of air pollution and mortality, such as was done in the National Morbidity, Mortality and Air Pollution study (NMMAPS), without excluding consideration of meta-analyses of previously published studies. Emphasis should also be placed on risk estimates obtained from analyzing data on multiple days so as to include delayed acute effects. (see Chapter 4). Such health-benefits estimates should be accompanied by a broad array of analyses of uncertainty, while at the same time understanding that a zero value is unlikely. In light of this recommended approach, future RIAs should give little or no weight to the assumption that there is no causal association between estimated reductions in the incidence of premature mortality and reduced ozone exposure.

Effect of Ozone Exposure on Life Expectancy

The impact of pollution on mortality may be quantified in terms of changes in mortality rates, number of deaths in a given period, or the months (or years) of life lost (or saved). EPA has calculated numbers of deaths prevented in a given year for each incremental reduction in PM or ozone concentration by applying the relative risk from epidemiologic studies to a baseline mortality rate. There is also interest in assessing the number of life-years saved, especially for cost-effectiveness analysis. The committee was asked whether there is an adequate basis for quantitatively characterizing the likely impact of reductions in short-term daily exposures to ozone on life expectancy.

Questions about the population at risk for mortality from short-term ozone exposure underlie most of the issues raised about interpreting epidemiologic time-series results in benefits assessment, including issues of susceptibility and short-term mortality displacement. Those issues are important for economic valuation. It is important to know that the willingness-to-pay (WTP)[4] values used are appropriate to the at-risk population, and this remains one of the key

[4]WTP is an estimate of the amount that a person is willing to pay for changing his or her mortality risk in a given period by a small amount (see Chapters 2 and 5).

uncertainties that the committee has highlighted in its report. For example, if we know from the cause of death associated with ozone exposure that the mortality risk is primarily for respiratory causes of death, we know that the age distribution of the at-risk population is skewed more toward the elderly, especially those experiencing cardiopulmonary compromise, than is total mortality in the general population. Thus, the WTP estimates should be relevant for a population with a higher average age. The currently available information lacks that level of detail. Although cardiorespiratory mortality stays at the center of the discussion, it is not known whether the age and risk profile of the people whose deaths are attributable to ozone corresponds to the typical person who dies of cardiorespiratory problems. The only information available is the population-average change in mortality rates, which translates into the population-level change in life expectancy.

A unified survival model of a dynamic cohort study that combines long-term and short-term air-pollution exposure has been developed (see Chapter 4). By incorporating life-table methods based on time-series risk estimates, the model can be used to estimate both life-years lost and number of additional deaths expected to occur in a specified period because of changes in air-pollution concentrations under restrictive assumptions. Further complexity can be gained by incorporating risks specific to age-sex groups. It is not clear, however, how long a period of long-term increased exposure is required to shorten life expectancy; it may be years. The time distribution of effect on longevity associated with immediate reductions in pollution also is not known. The risks based on time-series studies assume (by design) that reductions in pollution have an immediate benefit for longevity. As mentioned above, if the hazard and risk distributions in the population are not independent, estimates of excess deaths and life-years lost in the entire population attributable to changes in ozone exposure based on time-series studies will not equal the average of these quantities over all risk groups.

The most complete assessment of the association between acute exposure and death originates in distributed-lag models that integrate the distribution of the time between exposure and death. However, both the usual time-series model and the distributed-lag models focus on a short window of exposure. **Effects of long-term cumulated exposure are, by design, ignored in those studies. Reductions in life expectancy based on the time-series results can be calculated by using life-table methods, but lack of information about the at-risk population necessitates making the assumption that remaining life expectancy is comparable with of others in the same age-sex cohort.**

Use of Information from Acute-Effects and Chronic-Effects Studies

Evidence of long-term effects of ozone on mortality (or survival time) is presently weak; thus, the derivation of a relative risk to describe the association is more difficult than in the case of PM (see Chapter 4). The translation of the results into numbers of attributable deaths is possible but leaves two main challenges: First, the assessment of acute effects would be incomplete if based solely on the usual time-series studies, given that subacute (delayed) effects are not captured with these methods; only one study has published estimates of subacute effects of ozone based on distributed-lag models (Zanobetti and Schwartz 2008). Second, in the absence of abundant quantifiable evidence of chronic effects of ozone on mortality, total life-years lost because of both acute and chronic effects cannot be estimated with confidence. As mentioned previously, the current evidence from cohort studies, although it is weak, supports the notion that the effects estimates seen in the-series studies reflect only a portion of the total effect.

The question of how to use the epidemiologic information to translate into risk changes and benefits assessment is the subject of debate. So far, many risk assessments have derived as an intermediate step the number of deaths attributable to air pollution on an annual basis. That approach has several limitations that are of particular concern for chronic effects but, conceptually, apply also to acute effects.

A problem in discussing lives saved is related to the misleading message that any policy may prevent death whereas the only result one can expect is a postponement of death—a longer life. The term *attributable death* is at least less misleading than *lives saved*. Conceptually, however, *attributable death* still implies a body count rather than the life-years in premature deaths. In the case of death due to short-term ozone exposure of people with some pre-existing disease (either acute or chronic), the presentation of "annual attributable death" may be an acceptable approach. It should be noted, however, that in the case of quantifying chronic effects of pollution, assumed to complicate the underlying chronic diseases that increase the susceptibility to premature death, the attributable-death concept is less appropriate because it does not take fully into account the long-term dynamic in a population in which the underlying risk profiles change. A reduction in air pollution will lead to longer life expectancy and thus increase the number of elderly people. Age-adjusted mortality rates are expected to be lower under cleaner conditions, but the absolute number of deaths will steadily increase as the population ages. Consequently, the "attributable deaths" will not be the same throughout the years after an improvement in air quality (Hurley et al. 2000; Miller and Hurley 2003; Rabl 2006; Brunekreff et al. 2007).

> **Recommendation:** EPA should study emerging understanding of ozone-mortality associations and evaluate their use and implications in benefits assessments, including relationships between changes in mortality rates, annual deaths prevented, and years of life saved. The alternative approaches for expressing ozone mortality effects will lead to rather similar results if one is supposed to express only the most immediate (acute) effects of pollution changes. However, with the integration of subacute effects estimates, and particularly in the case of use of estimates of long-term chronic effects, the discrepancies between the approaches increase

and the conceptual flaws of the attributable-cases model become more pronounced.

VALUATION

An important question for benefits valuation is whether a person's value for reducing risks of death should be based on the magnitude of the life extension (that is, the loss of life-years that is avoided or prevented). Although there are intuitive reasons for saying yes, the question is how it should be done. Estimates of the value of a statistical life (VSL) are often used in cost-benefit analyses for programs expected to reduce mortality risks in a population. VSL estimates are derived from estimates of people's WTP for reducing their mortality risk in a given period by a small amount (see Chapters 2 and 5). If the effects of reducing mortality risk are measured in terms of years of life extension (increased remaining life expectancy), then it may be more appropriate to use a monetary value per statistical life-year (VSLY).

For any given risk reduction the average WTP value can be summarized as an average VSL or as an average VSLY if the remaining life expectancy and rate of time preference are known for the population from which the average WTP value was derived. Difficulties arise when either of these summary measures is assumed to be constant and used to estimate values for risk reductions in other populations or other risk reduction contexts.

EPA's current primary approach to economic valuation of mortality-risk reductions is a variation on its long-standing approach of using the same VSL for all annual mortality reductions. In its primary benefits estimates, EPA applies the VSL to all lives saved regardless of the age or health status of the population experiencing the change in mortality risk and regardless of the cause of the change.

Because there is some expectation that WTP for mortality-risk reduction will vary with the characteristics of the population affected or with the context of the risk change, EPA asked the committee to assess scientific approaches to assigning economic values to reductions in mortality risk associated with ozone reductions and to address the questions of economic valuation for different changes in life expectancy.

Valuation in the Context of Cost-Benefit Analysis

The charge to this committee concerns monetary valuation of mortality risk reduction for regulatory impact analyses that are based on the basic premises of cost-benefit analysis. In this context, therefore, we focus on WTP values for mortality risk reduction. Cost-benefit analysis, however, focuses on economic efficiency and many other ethical and legal factors are appropriate to consider in policy and regulatory decisions. In general, these issues should be considered separate from the cost-benefit analysis rather than interjected into the valuation estimates because this endangers the neutrality of the analysis. However, there may be some instances when such interjection is appropriate. For example, the decision by EPA to not adjust WTP estimates for local differences in income levels is justified because to do so creates a situation that favors greater environmental protection in wealthier locations, an outcome policy makers judge to be unfair and in many cases illegal.

The committee stresses that government decision-makers need information on how the WTP for mortality risk reductions varies by risk characteristics, population characteristics, and for both risk as a private good and as a public good. How they choose to use this information, however, is not strictly a technical decision, but depends on ethical precepts, legal precedent, the quality of the evidence and other factors that may be beyond the analysts' control or purview.

> **Recommendation:** We recommend that the issues contained in this finding of the committee be considered within the Office of Management and Budget and the agencies that use monetary values for mortality risk reductions in their regulatory analyses. These agencies should develop a plan for generating the information needed to determine how WTP varies for different populations and different risk contexts. In addition, there should be an exploration to determine the appropriate uses of this information. Such an exploration should go beyond economic considerations and include ethical and public policy perspectives.

Willingness-to-Pay Estimates

Both economic theory and the available empirical evidence are inconclusive about how individuals' WTP values for reducing their own risk vary with two important individual characteristics: age as a proxy for remaining life expectancy and health status. **We conclude that the empirical evidence is insufficient to support a specific quantitative adjustment of the WTP estimates to account for differences in remaining life expectancy, but we do not reject the concept that such adjustments may be appropriate.** It is plausible that people with shorter remaining life expectancy are willing to devote less of their resources to reducing their mortality risk than would people with longer remaining life expectancy.

Characteristics of the risk that may affect individual WTP values include the type of risk (for example, illness or accident) and its latency. The literature is inconclusive about the influence of risk characteristics (see below). The effects of latency on WTP values are straightforward conceptually (WTP for a future mortality-risk reduction should be less than that for an equivalent immediate risk reduction), and a small body of empirical literature supports such a concept. However, the epidemiologic literature is not sufficient to estimate the degree of latency of changes in mortality risk from ozone exposure.

> **Recommendation:** Although there are many concerns about the accuracy of using the same WTP estimate (or range of estimates) for all mortality-risk reductions, we recommend that EPA use this approach, with appropriate scaling to the size of the risk change, as the most scientifically supportable approach for monetary valuation of ozone-related mortality, given the currently available information in the economics and epidemiology literatures. Empirical evidence of how WTP varies with population or risk characteristics is not sufficiently consistent to support a change in this practice that EPA has been using for many years.

Estimates of Value of a Statistical Life-Year

The use of a constant VSLY in the valuation of increases in life expectancy requires the assumption that WTP values for mortality risk reductions be consistently declining with increasing age. **Available empirical evidence does not support that assumption and therefore does not support the use of a *constant* VSLY. The literature does not reject the use of a *non-constant* VSLY, however, and the epidemiological literature may favor reporting life-years saved.**

> **Recommendation:** Unless future research produces empirical support for the assumptions that underlie a constant VSLY, EPA should not attempt to make adjustments for remaining life expectancy by calculating life-years saved and using a constant VSLY to value them. It may be appropriate to calculate and report life-years saved (in addition to reporting changes in annual mortality ranges and reductions in premature deaths), but it is not appropriate to use a constant VSLY as a monetary valuation of life-years saved, except, perhaps in a bounding exercise. The committee cautions against use of such an analysis in anything but a sensitivity analysis, however (see below). There is likely to be good reason to use a non-constant VSLY or a non-constant VSL, once the literature is sufficient to make this transition. The committee stresses, however, that the status quo of using a uniform VSL should be continued until there is sufficient empirical evidence of how WTP for mortality risk reduction varies with differences in remaining life expectancy and other factors, which the committee concludes is not yet available.

Value of a Statistical Life, Individual Characteristics, and Risk Contexts

Most of the revealed-preference and stated-preference studies relied on by EPA to obtain estimates of the VSL have estimated WTP for mortality-risk reduction in a context (such as traffic accidents and workplace accidents) and for a population that differ from the context and population relevant to the pollution-related risks that EPA is assessing. **Applying the available estimates in EPA's**

assessments in a different risk context (illness vs accident) and for different population characteristics introduces considerable uncertainty about how these factors affect average WTP values. However, the current literature is inconclusive as to how much the WTP values may vary with these factors.

> **Recommendation:** EPA should ensure that the average WTP estimates selected from the literature reflect results of both revealed-preference and stated-preference studies. The agency should take into account the strengths and weaknesses of each study approach and consider how closely the available studies match the policy context in population at risk and type of risk. EPA should give less weight to wage-risk studies in selecting WTP estimates than it has in the past. Given the limited studies available for different risk contexts, it is difficult to say how much the WTP values may differ, but the wage-risk studies are a poor match to the population and to the risk context for the ozone-mortality case.

Sensitivity Analyses

The direction of the expected error, if any, in using the average VSL in the literature for valuing changes in ozone-related mortality risk is more likely to be toward overstating the WTP to reduce the risk. That is because greater mortality risk associated with ozone appears to be in the elderly population and because latent risk may be involved. Given this population's substantially less than average remaining life expectancy, it is possible that its WTP to reduce mortality risk would be less than the average WTP for the population as a whole. However, we expect that the lower WTP as a result of the elderly population's lower remaining life expectancy is offset to some extent by higher WTP because of poorer health status or higher baseline risk. Although results in the empirical literature are not consistent, several studies suggest that WTP to reduce mortality risk is constant or declines slightly with age. **That evidence suggests that, for ozone, a proportional adjustment of the VSL for remaining life expectancy (that is, using a constant VSLY) would result in WTP that is too low.**

> **Recommendation:** Given the uncertainty in the accuracy of available VSL estimates for ozone-related mortality, EPA should conduct sensitivity analyses with alternative estimates or assumptions. The purpose of sensitivity analyses is to see whether the overall conclusions of the cost-benefit comparison are changed—for example, whether net benefits are still positive under alternative economic-valuation assumptions. In general, there is less confidence in the sensitivity analyses because the alternative assumptions are more speculative than the primary assumptions or deviate from the status quo, which the committee feels puts a burden of proof on those who would overturn it. Therefore, EPA should present results in a way that does not give equal weight to the outcomes of sensitivity analyses and

the outcomes of the use of primary assumptions. However, results of sensitivity analyses can be included in summary conclusions. EPA should select alternative assumptions for sensitivity analyses on the basis of either theory or evidence. For example, different published empirical estimates of the relationship between WTP for mortality-risk reduction and age could be selected as illustrative of the range of results in the literature, including estimates of VSLY or VSL that vary with age.

RESEARCH RECOMMENDATIONS

The committee was asked to identify major gaps in knowledge about the benefits of reducing ozone exposure and the most promising research strategies to close the gaps, including additional data, analyses, or research needed to separate the contributions of ozone and other gaseous or particulate components of air pollution to the total short-term effect on premature mortality documented in the literature. The health-related recommendations (from Chapter 3 and 4) and the future research needs for valuation (from Chapter 5) briefly summarized in this section should be addressed as part of EPA's research strategy for estimating the benefits of reducing ambient ozone.

The committee recognizes that many of the recommended research activities are complex and will be difficult to undertake, and that sufficient resources may not be available to undertake all of them in the near term. Therefore, EPA and other agencies that might carry out the recommended research will need to set priorities and develop a strategy for addressing the various information needs.

Recommendations for Future Health Research

The health-related recommendations are presented in three broad categories related to enhancing exposure assessment, enhancing epidemiologic studies, and reducing uncertainty.

Enhanced Exposure Assessment for Epidemiology

Evaluate exposure metrics to determine whether and how much daily peak exposures, such as 1h or 8h exposures, and longer-term average exposures, such as over 24h, are associated with ozone-related mortality. Identify the appropriate exposure metrics to relate how control programs will affect ozone concentrations and health. Consider that control strategies may have quite different effects on 24-h average concentrations and shorter-term exposures. Also consider that peak short-term concentrations on lower-ozone days will respond differently from peak short-term concentrations on high-ozone days and often in the opposite direction.

Investigate potential confounding by evaluating regional and seasonal associations between ambient concentrations and exposures to ozone and $PM_{2.5}$ (and its components), how the data affect researchers' ability to control for confounding, and correlations between the various pollutant concentrations. When possible, focus on groups of individuals who are sensitive to ozone exposures and use data on the chemical and physical components and size distribution of $PM_{2.5}$.

Include PM speciation data to account for seasonal and geographic variability in the relationship between ozone and its potential confounders. Include the growing STN database in analyses of potential confounding of the ozone associations. Ensure that STN collects data frequently enough on the particle components most relevant to understanding the potential for confounding.

Monitor ozone in winter and report the measurements. The size of the winter program should be sufficient to allow researchers to examine seasonal differences in risk, how the differences vary spatially between communities with warmer and cooler winters, and ozone-mortality relationships at lower ozone concentrations. Ozone is a regional pollutant, and winter measurements need not be collected at all the summer locations; but collect them at the frequency of summer measurements.

Evaluate air-quality numerical models, such as the Community Multiscale Air Quality (CMAQ) model for use in ozone epidemiologic studies to extend the spatial scale of available data. Consider the uncertainty associated with the models before drawing inferences about mortality risk assessment associated with ozone exposure.

Evaluate ozone exposure models such as the Air Pollutants Exposure (APEX) model that are used to improve characterization of ozone exposure at the population level by taking into account human activity. Assess whether the models can be used to improve epidemiologic studies or benefits assessments.

Enhanced Epidemiologic Studies

Explore thresholds by studying panels of individuals considered to be susceptible to premature death from ozone exposure, such as those with impaired lung and heart function. Further explore how individual thresholds may vary and the extent to which thresholds depend on the frailty of the individual at any given moment.

Because it is not clear whether ozone is associated with mortality in the cooler months, examine warmer months separately. Conduct a sensitivity analysis on concentration-response relationships to capture more fully the uncertainties contributed by reliance on average fixed-site monitoring data to estimate population-average personal exposure.

Explore short-term mortality displacement and include use of alternative study methods, such as investigation of subjects who have diseases (for example, diabetes or heart disease) that are known to induce high mortality risk associated with air pollution.

Identify susceptibility characteristics that may have important effects on ozone-mortality relationships and develop a distribution of the estimates of the ozone mortality effect among the categories of susceptibility. To the extent that data are not available, use models and assumptions for sensitivity analysis.

Conduct distributed-lag analyses as one part of future epidemiologic investigations to improve understanding of the statistical distribution of time between an increase in the ambient concentration of ozone and a pattern of occurrence of death.

Conduct cohort studies to examine further the association between long-term ozone exposure and mortality. Develop long-term ozone-exposure models that can distinguish variations in exposure at the individual level and between and within cities. To the extent that new cohort studies strengthen evidence of long-term effects, consider including estimates based on that evidence in benefits assessments.

Study cardiovascular effects of ozone exposure, both in human and animal models. Design studies to identify genetic susceptibility factors.

Uncertainty

Characterize uncertainty of results of epidemiologic models and discuss their reliability and estimated uncertainty about which model (if any) is reasonably correct.

Consider Bayesian approaches for uncertainty analysis, including the possibility of additional expert elicitation once the recent experience with this approach to PM risk assessments has been evaluated.

Conduct sensitivity analysis intermittently as computational models and input distributions are developed to focus resources on the most important inputs or parts of the model.

Identify the bases of estimates to distinguish between data-derived estimates of some components (such as the concentration-response function) and the expert opinions about other components that are not supported by scientific data.

Recommendations for Future Research on Valuation

Explore how WTP varies with mortality-risk changes and changes in life expectancy, and explore how WTP varies with population characteristics such as age, health status, and baseline risk levels. Although the research needs would be addressed primarily by using stated-preference methods, explore ways to address research needs with further developments of revealed-preference methods.

Seek results on total age effects to understand better how age and remaining life expectancy affect WTP for reductions in mortality risks or extension of life expectancy. Ask that future (or even previous) studies report total age effects (that is, WTP by age cohort) in addition to effects of age alone on WTP for a small reduction in mortality risk in a given period.

Make datasets available for meta-analysis by urging researchers who receive EPA funding to make their datasets available for meta-analyses in addition to providing their published results.

Explore and develop full life-cycle methods for communicating and valuing changes in mortality risk as a shift in the survival curve, which plots survival probabilities in all future periods and from which life expectancy is derived.

Explore effects of mortality-risk characteristics on the valuation of reduction in these risks. Consider different types of risk (such as accident and illness) and the latency of the change in risk. Compare values of (or nonmonetary preferences for) risks in different contexts (e.g., Magat et al. 1996; DeShazo and Cameron 2004).

Explore the potential usefulness of studies that analyze preferences regarding public goods by developing approaches to distinguish paternalistic from nonpaternalistic altruism to avoid double-counting of benefits. Conduct conceptual analyses to learn how such results might be appropriately used in cost-benefit analysis, in which the usual paradigm is to sum beneficiaries' private WTP.

REGULATORY IMPACT ANALYSES INVOLVING OZONE MORTALITY

As discussed above and in Chapter 4, future RIAs based on currently available studies concerning ozone-related mortality should give greatest weight to the results of multicity time-series analyses. As EPA (2008b) had done for its RIA for the finalized ozone NAAQS, most of the emphasis should be placed on estimates based on systematic new multicity analyses, such as was done in the NMMAPS, without excluding consideration of meta-analyses of previously published studies. Future RIAs should incorporate research results on the mortality effects of chronic ozone exposure and research that addresses key uncertainties related to potential confounding factors, exposure measures, and susceptibility as appropriate.

Health-benefits estimates should be accompanied by a broad array of analyses of uncertainty, while at the same time understanding that a zero value would be unlikely. Future RIAs should give little or no weight to the assumption that there is no causal association between estimated reductions in the incidence of premature mortality and reduced ozone exposure unless new information emerges that refutes the interpretation of this association as causal. Presentations like that included in Table 7-14 of EPA's recent RIA for the finalized ozone

NAAQS, showing a variety of assumptions about the association between ozone exposure and mortality, should be revised in light of this recommendation (see Appendix B).

Risk analyses conducted with distributed-lag models over several days appear to capture the acute and the somewhat delayed subacute mortality effects of ozone exposure better than single-day data. Those models should be part of future benefits assessments to the extent that they are available.

There are many concerns about its accuracy for all mortality-risk reductions, but using a specific WTP value and a corresponding VSL estimate (or range of estimates) is the most scientifically supportable approach at this time to monetary valuation of ozone-related mortality, given the information in the economics literature. Before recommending a substantial change in EPA's approach for valuation of mortality risk reductions, it is necessary that there be fairly conclusive empirical evidence to support a specific change in the approach. It is the committee's judgment that the available evidence is not sufficient to support such a change at this time, but alternative approaches should be explored in sensitivity analyses and further research should be conducted to answer the questions raised about the validity of EPA's current approach. As new information emerges on population characteristics of those susceptible to mortality from ozone and on variations in WTP for mortality-risk reductions (or increases in life expectancy) due to different population characteristics, benefits-assessment methods may need to be revised.

EPA should consider placing greater emphasis on reporting decreases in age-specfic death rates and increases in life expectancy than on reporting estimates of lives saved. Such a change is needed to be responsive to recent reports that it is not possible to identify specific deaths attributable to air-pollution exposures (e.g., Brunekreef et al. 2007). For example, if the relative risk is 1.002 for a unit change in pollution (for example, for a change in tons of a pollutant emitted over a period), it means that there is about a 0.2% higher mortality rate for every unit increase in pollution in the population group to which the estimated relative risk applies. The relative risk is a proportional risk estimate. It may be estimated for the general population, but it would be preferable if it were estimated for the specific population groups that may be expected to have different susceptibilities to the effects of pollution exposure. In any case, it should be applied in a benefits assessment to the population for which it was estimated. If the scenario is for a two-unit reduction in pollution with a relative risk of 1.002 per unit change in pollution, and the relevant population has a baseline annual mortality rate of 1 in 100, the reduction in the annual mortality rate attributable to the change in pollution is about 4 per 100,000. For a population of 1 million, EPA would report that as 40 lives saved. The committee recommends that EPA report preferentially the annual mortality-rate change of 4 in 100,000 for the applicable population and to develop models for consistent calculation of life-expectancy changes and changes in numbers of deaths (e.g., Rabl 2006).

The WTP studies from which VSL estimates are derived match the annual mortality-rate change estimates. For example, the valuation literature reports

WTP estimates of $40-400 for a 4-in-100,000 risk reduction (translating to a VSL of $1-10 million). Turning the annual mortality-rate changes and WTP values into saved lives and values per life saved is a mathematical transformation that seems convenient for summary tables and policy assessments, but can be misleading and can obscure the underlying derivation and appropriate interpretation of the studies. It also ignores the effect of life-expectancy changes on the population composition over time, and this will cause the annual number of deaths to decrease at first and then to return to its previous level or even increase over time (e.g., Miller and Hurley 2003).

The health-related research and valuation research recommended in this report should be addressed as part of EPA's research strategy for estimating the mortality risk-reduction benefits of reducing exposure to ambient ozone. However, the research needs should not be viewed as a basis for postponing consideration of ozone mortality relationships in benefits assessment until more information is obtained. Also, it would be a mistake to assume that the committee's discussion of uncertainties and research needs broadly applies to the current understanding of air pollution and health in general. For example, this report has indicated where there is a greater understanding of many aspects of PM-mortality relationships relative to those for ozone. Continued enhancement of knowledge and methods for valuation of ozone mortality risk reduction benefits will inform future regulatory decision making and help in understanding the relative importance and value of effects caused by various pollutants.

References

Abbey, D.E., N. Nishino, W.F. McDonnell, R.J. Burchette, S.F. Knutsen, B.W. Lawrence, and J.X. Yang. 1999. Long-term inhalable particles and other air pollutants related to mortality in nonsmokers. Am. J. Respir. Crit. Care Med. 159(2):373-382.

Adams, W.C. 2002. Comparison of chamber and face-mask 6.6-hour exposures to ozone on pulmonary function and symptoms responses. Inhal. Toxicol. 14(7):745-764.

Adams, W.C. 2003. Comparison of chamber and face mask 6.6-hour exposure to 0.08 ppm ozone via square-wave and triangular profiles on pulmonary responses. Inhal. Toxicol. 15(3):265-281.

Adams, W.C. 2006. Comparison of chamber 6.6-h exposures to 0.04-0.08 ppm ozone via square-wave and triangular profiles on pulmonary responses. Inhal. Toxicol. 18(2):127-136.

Adler, M.D., and E.A. Posner. 2006. New Foundations of Cost-Benefit Analysis. Cambridge, MA: Harvard University Press.

Alberini, A. 2005. What is a life worth? Robustness of VSL values from contingent valuation surveys. Risk Anal. 25(4):783-800.

Alberini, A., M.L. Cropper, A. Krupnick, and N.B. Simon. 2004. Does the value of a statistical life vary with age and health status? Evidence from the U.S. and Canada. J. Environ. Econ. Manage. 48(1):769-792.

Alberini, A., M.L. Cropper, A. Krupnick, and N.B. Simon. 2006a. Willingness to pay for mortality risk reductions: Does latency matter? J. Risk Uncertain. 32(3):231-245.

Alberini, A., A. Hunt, and A. Markandya. 2006b. Willingness to pay to reduce mortality risks: Evidence from a Three-Country Contingent Valuation Study. Environ. Resour. Econ. 33(2):251-264.

Aldy, J.E., and W.K. Viscusi. 2007. Age differences in the value of a statistical life: Revealed preference evidence. Rev. Environ. Econ. Policy 1(2):241-260.

Aldy, J.E., and W.K. Viscusi. In press. Adjusting the value of a statistical life for age and cohort effects. Review of Economics and Statistics.

Allen, E., B.J. Becker, J.A. Berlin, S.C. Morton, I. Olkin, D. Rindskopf, A.R. Sampson, and D.B. Wilson. 2006. Report of the EPA Work Group on VSL Meta-Analyses. EE-0494. U.S. Environmental Protection Agency, National Center for Environmental Economics. July 25, 2006.

Anderson, H.R., R.W. Atkinson, J.L. Peacock, L. Marston, and K. Konstantinou. 2004. Meta-Analysis of Time-Series and Panel Studies of Particulate Matter and Ozone. World Health Organization [online]. Available: http://www.euro.who.int/document/e82792.pdf [accessed Dec. 5, 2007].

Aris, R.M., D. Christian, P.Q. Hearne, K. Kerr, W.E. Finkbeiner, and J.R. Balmes. 1993. Ozone-induced airway inflammation in human subjects as determined by airway lavage and biopsy. Am. Rev. Respir. Dis. 148(5):1363-1372.

Arjomandi, M., A.Witten, E. Abbritti, K. Reintjes, I. Schmidlin, W. Zhai, C. Solomon, and J. Balmes. 2005. Repeated exposure to ozone increases alveolar macrophage recruitment into asthmatic airways. Am. J. Respir. Crit. Care Med. 172(4):427-432.

Avol, E.L., W.C. Navidi, and S.D. Colome. 1998. Modeling ozone levels in and around Southern California homes. Environ. Sci. Technol. 32(4):463-468.

Bachmann, J. 2007. Will the circle be unbroken: A history of the U.S. National Ambient Air Quality Standards. J. Air Waste Manage. Assoc. 57(6):652-697.

Backus-Hazzard, G.S., R. Howden, and S.R. Kleeberger. 2004. Genetic susceptibility to ozone-induced lung inflammation in animal models of asthma. Curr. Opin. Allergy Clin. Immunol. 4(5):349-353.

Balmes, J.R., L.L. Chen, C. Scannell, I. Tager, D. Christian, P.Q. Hearne, T. Kelly, and R.M. Aris. 1996. Ozone-induced decrements in FEV1 and FVC do not correlate with measures of inflammation. Am. J. Respir. Crit. Care Med. 153(3):904-909.

Bayram, H., R.J. Sapsford, M.M. Abdelaziz, and O.A. Khair. 2001. Effect of ozone and nitrogen dioxide on the release of proinflammatory mediators from bronchial epithelial cells of nonatopic nonasthmatic subjects and atopic asthmatic patients in vitro. J. Allergy Clin. Immunol. 107(2):287-294.

Bell, M.L. 2007. Recent Evidence on the Relationship between Ozone and Mortality. Presentation at the 1st Meeting on Mortality Risk Reduction Benefits from Decreasing Tropospheric Ozone Exposure, March 29, 2007, Washington, DC.

Bell, M.L., and F. Dominici. 2008. Effect modification by community characteristics on the short-term effects of ozone exposure and mortality in 98 U.S. communities. Am. J. Epidemiol. 167(8):986-997.

Bell, M.L., A. McDermott, S.L. Zeger, J.M. Samet, and F. Dominici. 2004. Ozone and short-term mortality in 95 U.S. urban communities, 1987-2000. JAMA 292(19): 2372-2378.

Bell, M.L., F. Dominici, and J.M. Samet. 2005. A meta-analysis of time-series studies of ozone and mortality with comparison to the National Morbidity, Mortality and Air Pollution Study. Epidemiology 16(4):436-445.

Bell, M.L., R.D. Peng, and F. Dominici. 2006. The exposure-response curve for ozone and risk of mortality and the adequacy of current ozone regulations. Environ. Health Perspect. 114(4):532-536.

Bell, M.L., J.Y. Kim, and F. Dominici. 2007. Potential confounding of particulate matter on the short-term association between ozone and mortality in multisite time-series studies. Environ. Health Perspect. 115(11):1591-1595.

Bergamaschi, E., G. De Palma, P. Mozzoni, S. Vanni, M.V. Vettori, F. Broeckaert, A. Bernard, and A. Mutti. 2001. Polymorphism of quinone-metabolizing enzymes and susceptibility to ozone-induced acute effects. Am. J. Respir. Crit. Care Med. 163(6):1426-1431.

Bernstein, J.A., N. Alexis, C. Barnes, I.L. Bernstein, A. Nel, D. Peden, D. Diaz-Sanchez, S.M. Tarlo, and P.B. Williams. 2004. Health effects of air pollution. J. Allergy Clin. Immunol. 114(5):1116-1123.

Bickel, P., and R. Friedrich, eds. 2005. Extern E: Externalities of Energy: Methodology 2005 Update. Institut für Energiewirtschaft und Rationelle Energieanwendung – IER, Universität Stuttgart, Germany. EUR 2195. Luxembourg: Office for Official

Publications of the European Communities [online]. Available: http://www.externe.info/ [accessed Nov. 30, 2007].

Black, D.A., J. Galdo, and L. Lin. 2003. How Robust Are Hedonic Wage Estimates of the Price of Risk? Appendix C in Value of Statistical Life Analysis and Environmental Policy: A White paper for presentation to Science Advisory Board—Environmental Economics Advisory Committee. Report No. EE-0483. June 2003 [online]. Available: http://yosemite.epa.gov/ee/epa/eermfile.nsf/vwAN/EE-0483-04.pdf/$File/EE-0483-04.pdf [accessed Nov. 30, 2007].

Blomquist, G. 2004. Self-protection and averting behavior, values of statistical lives, and benefit cost analysis of environmental policy. Rev. Econ. Household 2(1):89-110.

Bosson, J., N. Stenfors, A. Bucht, R. Helleday, J. Pourazar, S.T. Holgate, F.J. Kelly, S. Wilson, A.J. Frew, and A. Blomberg. 2003. Ozone-induced bronchial epithelial cytokine expression differs between healthy and asthmatic subjects. Clin. Exp. Allergy 33(6):777-782.

Bosson, J., J. Pourazar, B. Forsberg, E. Adelroth, T. Sandström, and A. Blomberg. 2007. Ozone enhances the airway inflammation initiated by diesel exhaust. Respir. Med. 101(6):1140-1146.

Boylan, J.W., M.T. Odman, J.G. Wilkinson, and A.G. Russell. 2006. Integrated assessment modeling of atmospheric pollutants in the Southern Appalachian Mountains: Part II. Fine particulate matter and visibility. J. Air Waste Manage. Assoc. 56(1):12-22.

Brauer, M., P. Koutrakis, and J.D. Spengler. 1989. Personal exposure to acidic aerosols and gases. Environ. Sci. Technol. 23(11):1408-1412.

Brauer, M., M.W. Framp¬ton, P. Koutrakis, R.O. McClellan, W.F. McDonnell, S. Moolgavkar, D.W. North, A.E. Smith, R.L. Smith, and M.J. Utell. 2007. Critical Considerations in Evaluating Scientific Evidence of Health Effects of Ambient Ozone. Report of a Working Conference held in Rochester, NY, June 5-6, 2007.

Brook, R.D., J.R. Brook, B. Urch, R. Vincent, S. Rajagopalan, and F. Silverman. 2002. Inhalation of fine particulate air pollution and ozone causes acute arterial vasoconstriction in healthy adults. Circulation 105(13):1534-1536.

Brook, R.D., B. Franklin, W. Cascio, Y. Hong, G. Howard, M. Lipsett, R. Luepker, M. Mittleman, J. Samet, S.C. Smith, Jr., and I. Tager. 2004. Air pollution and cardiovascular disease: A statement for healthcare professionals from the Expert Panel on Population and Prevention Science of the American Heart Association. Circulation 109(21):2655-2671.

Brunekreef, B., B.G. Miller, and J.F. Hurley. 2007. The brave new world of lives sacrificed and saved, deaths attributed and avoided. Epidemiology 18(6):785-788.

Burnett, R.T., A. Dewanji, F. Dominic, M.S. Goldberg, A. Cohen, and D. Krewski. 2003. On the relationship between time series studies, dynamic population studies, and estimating loss of life due to short-term exposure to environmental risks. Environ. Health Perspect. 111(9):1170-1174.

Burnett, R.T., D. Stieb, J.R. Brook, S. Cakmak, R. Dales, M. Raizenne, R. Vincent, and T. Dann. 2004. Associations between short-term changes in nitrogen dioxide and mortality in Canadian cities. Arch. Environ. Health 50(5):228-236.

Cakmak, S., R.T. Burnett, and D. Krewski. 1999. Methods for detecting and estimating population threshold concentrations for air pollution-related mortality with exposure measurement error. Risk Anal. 19(3):487-496.

CalEPA (California Environmental Protection Agency). 2006. The Children's Health Study. Air Resources Board, California Environmental Protection Agency [online]. Available: http://www.arb.ca.gov/research/chs/chs.htm [accessed Nov. 6, 2007].

CalEPA (California Environmental Protection Agency). 2007. Review of the California Ambient Air Quality Standard for Ozone. Air Resources Board, California Environmental Protection Agency [online]. Available: http://www.arb.ca.gov/research/aaqs/ozone-rs/ozone-final/ozone-final.htm#Summary [accessed Jan. 2, 2008].

Carroll, R.J., D. Ruppert, and L.A. Stefanski. 1995. Measurement Error in Nonlinear Models. London: Chapman and Hall.

Champ, P.A., and R.C. Bishop. 2006. Is willingness to pay for a public good sensitive to the elicitation format? Land Econ. 82(2):162-173.

Chang, L.T., P. Koutrakis, P.J. Catalano, and H.H. Suh. 2000. Hourly personal exposures to fine particles and gaseous pollutants - results from Baltimore, Maryland. J. Air Waste Manage. Assoc. 50(7):1223-1235.

Chen, C.Y., A.C. Bonham, C.G. Plopper, and J.P. Joad. 2003. Neuroplasticity in nucleus tractus solitarius neurons after episodic ozone exposure in infant primates. J. Appl. Physiol. 94(2):819-827.

Chen, L.L., I.B. Tager, D.B. Peden, D.L. Christian, R.E. Ferrando, B.S. Welch, and J.R. Balmes. 2004. Effect of ozone exposure on airway responses to inhaled allergen in asthmatic subjects. Chest 125(6):2328-2335.

Chestnut, L.G., R.D. Rowe, and W.S. Breffle. 2004a. Economic Valuation of Mortality Risk Reduction: Stated Preference Approach in Canada. Report prepared for Health Canada under contracts 02-OPPS-102, H1021-02-0642/02 by Stratus Consulting Inc., Boulder, CO.

Chestnut, L.G., R.D. Rowe, L.K. Lazo, and W.S. Breffle. 2004b. Economic Valuation of Mortality Risk Reduction: Vol. 1. Stated Preference Approach. Report prepared for U.S. Environmental Protection Agency, Cooperative Agreement CR824393-01-0, by Stratus Consulting Inc., Boulder, CO. November 2004 [online]. Available: http://yosemite.epa.gov/ee/epa/eermfile.nsf/vwAN/EE-0488-01.pdf/$File/EE-0488-01.pdf [accessed Nov. 30, 2007].

Chow, J.C. 2003. Introduction to special topic: Weekend and weekday differences in ozone levels. J. Air Waste Manage. Assoc. 53(7):771.

Chuang, K.J., C.C. Chan, T.C. Su, C.T. Lee, and C.S. Tang. 2007. Urban air pollution on inflammation, oxidative stress, coagulation and autonomic dysfunction. Am. J. Respir. Crit. Care Med. 176(4):370-376.

Cohan, D.S., A. Hakami, Y. Hu, and A.G. Russell. 2005. Nonlinear response of ozone to emissions: Source apportionment and sensitivity analysis. Environ. Sci. Technol. 39(17):6739-6748.

Corradi, M., R. Alinovi, M. Goldoni, M. Vettori, G. Folesani, P. Mozzoni, S. Cavazzini, E. Bergamaschi, L. Rossi, and A. Mutti. 2002. Biomarkers of oxidative stress after controlled human exposure to ozone. Toxicol. Lett. 134(1-3):219-225.

Corso, P.S., J.K. Hammitt, and J.D. Graham. 2001. Valuing mortality-risk reductions: Using visual aids to improve the validity of contingent valuation. J. Risk Uncertain. 23(2):165-184.

Coyle, D., D. Stieb, D. Krewski, R.T. Burnett, P. DeCivita, D. Krewski, Y. Chen, and M.J. Thun. 2003. Impact of particulate air pollution on quality-adjusted life expectancy in Canada. J. Toxicol. Environ. Health A 66(16-19):1847-1863.

Cropper, M.L., and U. Subramanian. 1995. Public Choices between Lifesaving Programs: How Important Are Lives Saved? Policy Research Working Paper 1497. The World Bank, Washington, DC. August 1995[online]. Available: http://go.worldbank.org/XTQPY52SR0 [accessed Dec. 3, 2007].

Cropper, M.L., and F.G. Sussman. 1990. Valuing future risks to life. J. Environ. Econ. Manage. 19(2):160-174.

Cropper, M.L., S.K. Aydede, and P.R. Portney. 1994. Preferences for life saving programs: How the public discounts time and age. J. Risk Uncertain. 8(3):243-265.

David, G.L., I. Romieu, J.J. Sienra-Monge, W.J. Collins, M. Ramirez-Aquilar, B.E. del Rio-Navarro, N.I. Reyes-Ruiz, R.W. Morris, J.M. Marzec, and S.J. London. 2003. Nicotinamine adenine dinucleotide (phosphate) reduced: Quinone oxidoreductase and glutathione s-transferase M1 polymorphisms and childhood asthma. Am. J. Respir. Crit. Care Med. 168(10):1199-1204.

DEFRA (U.K. Department for Environment, Food and Rural Affairs). 2007. Air Quality Strategy for England, Scotland, Wales and Northern Ireland. U.K. Department for Environment, Food and Rural Affairs. July 17, 2007 [online]. Available: http://www.defra.gov.uk/environment/airquality/strategy/index.htm [accessed Jan. 2, 2007].

Depuydt, P.O., B.N. Lambrecht, G.F. Joos, and R.A. Pauwels. 2002. Effect of ozone exposure on allergic sensitization and airway inflammation induced by dendritic cells. Clin. Exp. Allergy 32(3):391-396.

Desaigues, B., A. Rabl, D. Ami, K.B. My, S. Masson, M.A. Salomon, and L. Santoni. 2004. Monetary Valuation of Air Pollution Mortality: Current Practice, Research Needs, and Lessons from a Contingent Valuation. Report Centre d'Energétique, Ecole des Mines de Paris. August 22, 2004 [online]. Available: http://www.czp.cuni.cz/ekonomie/Valuace%20papers/DESAIGUES-RABL%20Mortalita%20 current%20practise.pdf [accessed Nov. 30, 2007].

DeShazo, J.R., and T.A. Cameron. 2004. Mortality and Morbidity Risk Reduction: An Empirical Life Cycle Model of Demand with Two Types of Age Effects. Department of Policy Studies, University of California, Los Angeles [online]. Available: http://faculty.spa.ucla.edu/deshazo/pdf/14/Age_paper_allparts.pdf [accessed Nov. 30, 2007].

Detels, R., D.P. Tashkin, J.W. Sayre, S.N. Rokaw, A.H. Coulson, F.J. Massey Jr., and D.H. Wegman. 1987. The UCLA population studies of chronic obstructive respiratory disease. 9. Lung function changes associated with chronic exposure to photochemical oxidants; A cohort study among never-smokers. Chest 92(4):594-603.

Devlin, R.B., W.F. McDonnell, R. Mann, S. Becker, D.E. House, D. Schreinemachers, and H.S. Koren. 1991. Exposure of humans to ambient levels of ozone for 6.6 hours causes cellular and biochemical changes in the lung. Am. J. Respir. Cell Mol. Biol. 4(1):72-81.

Devlin, R.B., K.P. McKinnon, T. Noah, S. Becker, and H.S. Koren. 1994. Ozone-induced release of cytokines and fibronectin by alveolar macrophages and airway epithelial cells. Am. J. Physiol. 266(6 Pt 1):L612-L619.

Dickerson, R.R., S. Kondragunta, G. Stenchikov, K.L. Civerolo, B.G. Doddridge, and B.N. Holben. 1997. The impact of aerosols on solar ultraviolet radiation and photochemical smog. Science 278(5339):827-830.

Dockery, D.W., C.A. Pope, III, X. Xu, J.D. Spengler, J.H. Ware, M.E. Fay, B.G. Ferris, Jr., and F.E. Speizer. 1993. An association between air pollution and mortality in six U.S. cities. N. Engl. J. Med. 329(24):1753-1759.

Dockins, C., K. Maquire, N. Simon, and M. Sullivan. 2004. Value of Statistical Life Analysis and Environmental Policy: A White Paper for Presentation to the Science Advisory Board—Environmental Economics Advisory Committee. U.S. Environmental Protection Agency, Office of Policy, Economics and Innovation, National Center for Environmental Economics. April 21, 2004.

Dockins, C., K. Maguire, and N. Simon. 2006. Willingness to Pay for Environmental Health Risk Reductions When There are Varying Degrees of Life Expectancy: A White Paper. U.S. Environmental Protection Agency, Office of Policy, Economics and Innovation, National Center for Environmental Economics. September 1, 2006.

Dodge, M.C. 1977. Combined use of modeling techniques and smog chamber data to derive ozone-precursor relationships. Pp. 881-889 in International Conference on Photochemical Oxidant Pollution and Its Control- Proceedings, Vol. II, B. Dimitriades, ed. EPA/600/3-77-001b. U.S. Environmental Protection Agency, Research Triangle Park, NC.

Dolan, P., R. Shaw, A. Tsuchiya, and A. Williams. 2005. QALY maximization and people's preferences: A methodological review of the literature. Health Econ. 14(2):197-208.

Dominici, F., A. McDermott, S.L. Zeger, and J.M. Samet. 2003. Airborne particulate matter and mortality: Time-scale effects in four U.S. cities. Am. J. Epidemiol. 157(12):1055-1065 [with invited commentary].

DOT (U.S. Department of Transportation). 2002. Revised Departmental Guidance: Treatment of Value of Life and Injuries in Preparing Economic Evaluations. Memorandum to Assistant Secretaries and Modal Administrators from Kirk K. Van Tine, General Counsel, and Linda Lawson, Acting Deputy Assistant Secretary for Policy and International Affairs, U.S. Department of Transportation. January 29, 2002 [online]. Available: http://ostpxweb.ost.dot.gov/policy/Data/VSL02 guid.pdf [accessed Jan. 2, 2008].

Drazen, J.M., and D.R. Beier. 1997. Genetics of air pollution. Nat. Genet. 17(4):365-366

Drechsler-Parks, D.M. 1995. Cardiac output effects of O_3 and NO_2 exposure in healthy older adults. Toxicol. Ind. Heatlh 11(1):99-109.

Drechsler-Parks, D.M., J.F. Bedi, and S.M. Horvath. 1987. Pulmonary function responses of older men and women to ozone exposure. Exp. Gerontol. 22(2):91-101.

Drechsler-Parks, D.M., J.F. Bedi, and S.M. Horvath. 1989. Pulmonary function responses of young and older adults to mixtures of O_3, NO_2, and PAN. Toxicol. Ind. Health 5(3):505-517.

Dublin, L.I., and A.J. Lotka. 1930. The Money Value of a Man, 1st Ed. New York: Ronald Press.

Emmons, K.M., and W.M. Foster. 1991. Smoking cessation and acute airway response to ozone. Arch. Environ. Health 46(5):288-295.

EPA (U.S. Environmental Protection Agency). 1997a. Regulatory Impact Analyses for the Particulate Matter and Ozone: National Ambient Air Quality Standards and Proposed Regional Haze Rule. Prepared by Innovative Strategies and Economics Group, Office of Air Quality Planning and Standards, U.S. Environmental Protection Agency, Research Triangle Park, NC. July 17, 1997 [online]. Available: http://www.epa.gov/ttn/oarpg/naaqsfin/ria.html [accessed Nov. 14, 2007].

EPA (U.S. Environmental Protection Agency). 1997b. The Benefits and Costs of the Clean Air Act, 1970-1990. Prepared for U.S. Congress by U.S. Environmental Protection Agency. October 1997 [online]. Available: http://www.epa.gov/air/sect 812/copy.html [accessed Nov. 13, 2007].

EPA (U.S. Environmental Protection Agency). 1997c. Guiding Principles for Monte Carlo Analysis. EPA/630/R-97/001. Risk Assessment Forum, U.S. Environmental Protection Agency, Washington, DC. March 1997 [online]. Available: http://www. epa.gov/ncea/raf/montecar.pdf [accessed Dec. 4, 2007].

EPA (U.S. Environmental Protection Agency). 1999a. The Benefits and Costs of the Clean Air Act, 1990-2010. EPA Report to Congress. EPA-410-R-99-001. Office of Air and Radiation, Office of Policy, U.S. Environmental Protection Agency. November 1999 [online]. Available: http://www.epa.gov/air/sect812/1990-2010/full rept.pdf [accessed Nov. 13, 2007].

EPA (U.S. Environmental Protection Agency). 1999b. Final Tier 2 Rule: Air Quality Estimation, Selected Health and Welfare Benefits Methods, and Benefits Analysis Results. EPA-420-R-99-032. Office of Air Quality Planning and Standards, Research Triangle Park, NC. December [online]. Available: http://www.epa.gov/tier2/frm/tsd/r99032.pdf [accessed Dec. 3, 2007].

EPA (U.S. Environmental Protection Agency). 2000a. Guidelines for Preparing Economic Analysis. EPA 240-R-00-003. Office of Administrator, U.S. Environmental Protection Agency, Washington, DC. September 2000 [online]. Available: http://yosemite.epa.gov/ee/epa/eed.nsf/webpages/Guidelines.html [accessed Nov. 19, 2007].

EPA (U.S. Environmental Protection Agency). 2000b. Regulatory Impact Analysis: Heavy-Duty Engine and Vehicle Standards and Highway Diesel Fuel Sulfur Control Requirements. EPA420-R-00-026. Office of Air and Radiation, U.S. Environmental Protection Agency, Washington, DC. December 2000 [online]. Available: http://www.epa.gov/otaq/highway-diesel/regs/exec-sum.pdf [accessed Nov. 28, 2007].

EPA (U.S. Environmental Protection Agency). 2002. Final Regulatory Support Document: Control of Emissions from Unregulated Nonroad Engines. EPA 420-R-02-022. Office of Air and Regulation, U.S. Environmental Protection Agency. September 2002 [online]. Available: http://www.epa.gov/otaq/regs/nonroad/2002/r02022.pdf [accessed Jan. 2, 2008].

EPA (U.S. Environmental Protection Agency). 2003a. Benefits and Costs of the Clean Air Act 1990-2020: Revised Analytical Plan for EPA's Second Prospective Analysis. Office of Policy Analysis and Review, U.S. Environmental Protection Agency [online]. Available: http://www.epa.gov/air/sect812/1990-2020/mainbody51203.pdf [accessed Nov. 13, 2007].

EPA (U.S. Environmental Protection Agency). 2003b. Technical Addendum: Methodologies for the Benefit Analysis of the Clear Skies Act of 2003. U.S. Environmental Protection Agency. September 2003 [online]. Available: http://www.epa.gov/clearskies/tech_addendum.pdf [accessed Dec. 3, 2007].

EPA (U.S. Environmental Protection Agency). 2003c. National Air Quality and Emission Trends Report: 2003 Special Studies Edition. EPA 454/R-03-005. Office of air Quality and Standards, U.S. Environmental Protection Agency, Research Triangle Park, NC. September 2003 [online]. Available: http://www.epa.gov/air/airtrends/aqtrnd03/dl_graph.html [accessed Nov. 2, 2007].

EPA (U.S. Environmental Protection Agency). 2003d. Consolidated Human Activities Database (CHAD). U.S. Environmental Protection Agency [online]. Available: http://www.epa.gov/chadnet1/ [accessed Jan. 2, 2008].

EPA (U.S. Environmental Protection Agency). 2005a. Regulatory Impact Analysis for the Final Clean Air Interstate Rule. EPA-452/R-05-002. Air Quality Strategies and Standards Division, Emission, Monitoring, and Analysis Division and Clean Air Markets Division, Office of Air and Radiation, U.S. Environmental Protection Agency. March 2005 [online]. Available: http://www.epa.gov/cair/pdfs/final tech08.pdf [accessed Nov. 28, 2007].

EPA (U.S. Environmental Protection Agency). 2005b. Technical Support Document for the Final Clean Air Interstate Rule: Air Quality Modeling. Office of Air Quality Planning and Standards, U.S. Environmental Protection Agency, Research Triangle Park, NC. March 2005 [online]. Available: http://www.epa.gov/cair/pdfs/final tech02.pdf [accessed Mar. 11, 2008].

EPA (U.S. Environmental Protection Agency). 2005c. Ozone and Particle Pollution: CAIR, Together with Other Clean Air Programs, Will Bring Cleaner Air to Areas in the East-2020. Charts and Tables for Final Clean Air Interstate Rule [onluine]. Available: http://www.epa.gov/cair/charts_files/nonattain_maps.pdf [accessed Mar. 27, 2008].

EPA (U.S. Environmental Protection Agency). 2006a. Air Quality Criteria for Ozone and Related Photochemical Oxidants (Final). EPA/600/R-05/004aF-cF. Office of Research and Development, National Center for Environmental Assessment, U.S. Environmental Protection Agency, Research Triangle Park, NC. February 2006.

EPA (U.S. Environmental Protection Agency). 2006b. Clear Skies. Office of Air and Radiation, U.S. Environmental Protection Agency [online]. Available: http://www.epa.gov/air/clearskies/ [accessed Mar. 28, 2008].

EPA (U.S. Environmental Protection Agency). 2006c. Regulatory Impact Analysis for the 2006 National Ambient Air Quality Standards for Particle Pollution. Technology Transfer Network Economics and Cost Analysis Support, U.S. Environmental Protection Agency. October 6, 2006 [online]. Available: http://www.epa.gov/ttn/ecas/ria.html [accessed June 26, 2008].

EPA (U.S. Environmental Protection Agency). 2007a. Review of the National Ambient Air Quality Standards for Ozone: Policy Assessment of Scientific and Technical Information. OAQPS Staff Paper. EPA-452/R-07-003. Office of Air Quality Planning and Standards, U.S. Environmental Protection Agency, Research Triangle Park, NC. January 2007 [online]. Available: http://www.epa.gov/ttn/naaqs/standards/ozone/data/2007_01_ozone_staff_paper.pdf [accessed Oct. 31, 2007].

EPA (U.S. Environmental Protection Agency). 2007b. Regulatory Impact Analysis of the Proposed Revisions to the National Ambient Air Quality Standards for Ground-Level Ozone. EPA-452/R-07-008. Office of Air Quality Planning and Standards, U.S. Environmental Protection Agency, Research Triangle Park, NC. July 2007 [online]. Available: http://www.epa.gov/ttn/ecas/ria.html#ria2007 [accessed Oct. 31, 2007].

EPA (U.S. Environmental Protection Agency). 2007c. Review of the NAAQS Process. National Ambient Air Quality Standards, Office of Air and Radiation, U.S. Environmental Protection Agency [online]. Available: http://www.epa.gov/ttnmain1/naaqs/#review [accessed Jan 2, 2008].

EPA (U.S. Environmental Protection Agency). 2007d. Review of the National Ambient Air Quality Standards for Ozone: Policy Assessment of Scientific and Technical Information, Appendices to OAQPS Staff Paper. EPA 452/R-07/007. Office of Air Quality Planning and Standards, U.S. Environmental Protection Agency, Research Triangle Park, NC. July 2007 [online]. Available: http://www.epa.gov/ttn/naaqs/standards/ozone/data/2007_07_o3sp_appendices.pdf [accessed Nov. 5, 2007].

EPA (U.S. Environmental Protection Agency). 2007e. Proposal to Revise the National Ambient Air Quality Standards for Ground-level Ozone-Maps. U.S. Environmental Protection Agency Presentation June 21, 2007 [online]. Available: http://www.epa.gov/groundlevelozone/pdfs/20070621_maps.pdf [accessed Nov. 6, 2007].

EPA (U.S. Environmental Protection Agency). 2007f. Particulate Matter (PM)-National Ambient Air Quality Standards. U.S. Environmental Protection Agency [online]. Available: http://www.epa.gov/ttn/naaqs/standards/pm/s_pm_index.html [accessed Nov. 13, 2007].

EPA (U.S. Environmental Protection Agency). 2007g. Clean Diesel Trucks, Buses, and Fuel: Heavy-Duty Engine and Vehicle Standards and Highway Diesel Fuel Sulfur Control Requirements (the "2007 Heavy-Duty Highway Rule"). Office of Regulation and Standards, Office of Transportation and Air Quality, U.S. Environmental Protection Agency [online]. Available: http://www.epa.gov/otaq/highway-diesel/regs/2007-heavy-duty-highway.htm [accessed Jan. 2, 2008].

EPA (U.S. Environmental Protection Agency). 2007h. Clean Air Interstate Rule. Office of Air and Radiation, U.S. Environmental Protection Agency [online]. Available: http://www.epa.gov/cair/ [accessed Dec. 5, 2007].

EPA (U.S. Environmental Protection Agency). 2007i. Proposed Emission Standards for New Nonroad Spark-Ignition Engines, Equipment, and Vessels. EPA420-F-07-032. Office of Transportation and Air Quality, U.S. Environmental Protection Agency. April 2007 [online]. Available: http://www.epa.gov/otaq/regs/nonroad/marinesi-equipld/420f07032.htm [accessed Jan. 2, 2008].

EPA (U.S. Environmental Protection Agency). 2007k. Ozone Population Exposure Analysis for Selected Urban Areas. EPA-452/R-07-010. Office of Air Quality Planning and Standards, U.S. Environmental Protection Agency, Research Triangle Park, NC. July 2007 [online]. Available: http://www.epa.gov/ttn/naaqs/standards/ozone/data/2007_07_o3_exposure_tsd.pdf [accessed Mar. 11, 2008].

EPA (U.S. Environmental Protection Agency). 2008a. Ozone Standards. Regulatory Actions. Office of Air and Radiation, U.S. Environmental Protection Agency [online]. Available: http://www.epa.gov/groundlevelozone/actions.html [accessed Mar. 13, 2008].

EPA (U.S. Environmental Protection Agency). 2008b. Final Ozone NAAQS Regulatory Impact Analysis. Office of Air Quality Planning and Standards, U.S. Environmental Protection Agency, Research Triangle Park, NC. March 2008 [online]. Available: http://www.epa.gov/ttnecas1/ria.html [accessed Apr. 11, 2008].

EPA SAB (U.S. Environmental Protection Agency Science Advisory Board). 2007. SAB Advisory on EPA's Issues in Valuing Mortality Risk Reduction. EPA-SAB-08-001. Science Advisory Board, U.S. Environmental Protection Agency, Washington, DC. October 12, 2007 [online]. Available: http://www.epa.gov/sab/pdf/sab-08-001.pdf [accessed Nov. 2, 2007].

Evans, M.F., and V.K. Smith. 2006. Do we really understand the age-VSL relationship? Resour. Energy Econ. 28(3):242-261.

Fanucchi, M.V., C.G. Plopper, M.J. Evans, D.M. Hyde, L.S. Van Winkle, L.J. Gershwin, and E.S. Schelegle. 2006. Cyclic exposure to ozone alters distal airway development in infant rhesus monkeys. Am. J. Physiol. Lung Cell Mol. Physiol. 291(4):L644-L650.

Finkelstein, J.N., and C.J. Johnston. 2004. Enhanced sensitivity of the postnatal lung to environmental insults and oxidant stress. Pediatrics 113(Suppl. 4):1092-1206.

Finlayson-Pitts, B.J., and J.N. Pitts. 2000. Chemistry of the Upper and Lower Atmosphere: Theory, Experiments, and Applications. San Diego, CA: Academic Press.

Fiore, A., D.J. Jacob, H. Liu, R.M. Yantosca, T.D. Fairlie, and Q. Li. 2003. Variability in surface ozone background over the United States: Implications for air quality policy. J. Geophys. Res. 108(D24):4787 doi:10.1029/2003JD003855.

Flanagan, J.B., R.K. Jayanty, E.E. Rickman, and M.R. Peterson. 2006. PM2.5 speciation trends network: Evaluation of whole-system uncertainties using data from sites with collocated samplers. J. Air Waste Manage. Assoc. 56(4):492-499.

Folinsbee, L.J., S.M. Horvath, P.B. Raven, J.F. Bedi, A.R. Morton, B.L. Drinkwater, N.W. Bolduan, and J.A. Gliner. 1977. Influence of exercise and heat stress on pulmonary function during ozone exposure. J. Appl. Physiol. 43(3):409-413.

Folinsbee, L.J., J.F. Bedi, and S.M. Horvath. 1981. Combined effects of ozone and nitrogen dioxide on respiratory function in man. Am. Ind. Hyg. Assoc. J. 42(7):534-541.

Foster, W.M., J.A. Silver, and M.L. Groth. 1993. Exposure to ozone alters regional function and particle dosimetry in the human lung. J. Appl. Physiol. 75(5):1938-1945.

Foster, W.M., R.H. Brown, K. Macri, and C.S. Mitchell. 2000. Bronchial reactivity of healthy subjects: 18-20 h post-exposure to ozone. J. Appl. Physiol. 89(5):1804-1810.

Franklin, M., and J. Schwartz. 2008. The impact of secondary particles on the association between ambient ozone and mortality. Environ. Health Perspect. 116(4):453-458.

Freeman, A.M., III. 2003. The Measurement of Environmental and Resource Values: Theory and Methods, 2nd Ed. Washington, DC: Resources for the Future.

Fung, K., D. Krewski, R. Burnett, and F. Dominici. 2005. Testing the harvesting hypothesis by time-domain regression analysis. I: Baseline analysis. J. Toxicol. Environ. Health A 68(13-14):1137-1154.

Galizia, A., and P.L. Kinney. 1999. Long term residence in areas of high ozone: Associations with respiratory health in a nationwide sample of nonsmoking young adults. Environ Health Perspect. 107(8):675-679.

Gauderman, W.J., R. McConnell, F. Gilliland, S. London, D. Thomas, E. Avol, H. Vora, K. Berhane, E.B. Rappaport, F. Lurmann, H.G. Margolis, and J. Peters. 2000. Association between air pollution and lung function growth in southern California children. Am. J. Respir. Crit. Care Med. 162(4 Pt 1):1383-1390.

Gauderman, W.J., G.F. Gilliland, H. Vora, E. Avol, D. Stram, R. McConnell, D. Thomas, F. Lurmann, H.G. Margolis, E.B. Rappaport, K. Berhane, and J.M. Peters. 2002 Association between air pollution and lung function growth in southern California children: Results from a second cohort. Am. J. Respir. Crit. Care Med. 166(1):76-84.

Gauderman, W.J., E. Avol, F. Gilliland, H. Vora, D. Thomas, K. Berhane, R. McConnell, N. Kuenzli, F. Lurmann, E. Rappaport, H. Margolis, D. Bates, and J. Peters. 2004. The effect of air pollution on lung development from 10 to 18 years of age. N. Engl. J. Med. 351(11):1057-1067.

Gauderman, W.J., H. Vora, R. McConnell, K. Berhane, F. Gilliland, D. Thomas, F. Lurmann, E. Avol, N. Kunzli, M. Jerrett, and J. Peters. 2007. Effect of exposure to traffic on lung development from 10 to 18 years of age: A cohort study. Lancet 369(9561):571-577.

Gegax, D., S. Gerking, and W. Schulze. 1991. Perceived risk and the marginal value of safety. Rev. Econ. Stat. 73(4):589-596.

Gilks, W.R., S. Richardson, and D.J. Spiegelhalter. 1996. Markov Chain Monte Carlo in Practice. London: Chapman & Hall.

Gilliland, F.D., W.J. Gauderman, H. Vora, E. Rappaport, and L. Dubeau. 2002. Effects of glutathione-S-transferase M1, T1, and P1 on childhood lung function growth. Am. J. Respir. Crit. Care Med. 166(5):710-716.

Gold, D.R., G. Allen, A. Damokosh, P. Serrano, C. Hayes, and M. Castillejos. 1996. Comparison of outdoor and classroom ozone exposures for school children in Mexico City. J. Air Waste Manage. Assoc. 46(4):335-342.

Goldberg, M.S., R.T. Burnett, J.C. Bailar III, J. Brook, Y. Bonvalot, R. Tamblyn, R. Singh, M.F. Valois, and R. Vincent. 2001. The association between daily mortality and ambient air particle pollution in Montreal, Quebec. 2. Cause-specific mortality. Environ. Res. 86(1):26-36.

Goldberg, M.S., R.T. Burnett, J.F. Yale, M.F. Valois, and J.R. Brook. 2005. Associations between ambient air pollution and daily mortality among persons with diabetes and cardiovascular disease. Environ. Res. 100(2):255-267.

Gong, H., Jr., R. Wong, R.J. Sarma, W.S. Linn, E.D. Sullivan, D.A. Shamoo, K.R. Anderson, and S.B. Prasad. 1998. Cardiovascular effects of ozone exposure in human volunteers. Am. J. Respir. Crit. Care Med. 158(2):538-546.

Graham, J.D. 2003. Benefit-Cost Methods and Lifesaving Rules. Memorandum to the President's Management Council, from John D. Graham, Administrator, Office of Information and Regulatory Affairs, Office of Management and Budget, Washington, DC. May 30, 2003 [online]. Available: http://www.whitehouse.gov/omb/inforeg/pmc_benefit_cost_memo.pdf [accessed Nov. 19, 2007].

Gryparis, A., B. Forsberg, K. Katsouyanni, A. Analitis, G. Touloumi, J. Schwartz, E. Samoli, S. Medina, H.R. Anderson, E.M. Niciu, H.E. Wichmann, B. Kriz, M. Kosnik, J. Skorkovsky, J.M. Vonk, and Z. Dörtbudak. 2004. Acute effects of ozone on mortality from the "air pollution and health: A European approach" project. Am. J. Respir. Crit. Care Med. 170(10):1080-1087.

Guria, J., W. Jones, M.W. Jones-Lee, M. Keall, J. Leung, and G. Loomes. 1999. The Value of Statistical Life and Prevention of Injuries in New Zealand. Land Transport Safety Authority of New Zealand, Wellington, New Zealand.

Hammitt, J.K. 2002. QALYs versus WTP. Risk Anal. 22(5):985-1001.

Hammitt, J.K. 2007. Valuing changes in mortality risk: Lives saved versus life years saved. Rev. Environ. Econ. Policy 1(2):228-240.

Hammitt, J.K., and J.D. Graham. 1999. Willingness to pay for health protection: Inadequate sensitivity to probability? J. Risk Uncertain. 8(1):33-62.

Hammitt, J.K., and M.E. Ibarrarán. 2006. The economic value of reducing fatal and nonfatal occupational risks in Mexico City using actuarial- and perceived-risk estimates. Health Econ. 15(12):1329-1335.

Hammitt, J.K., and J.T. Liu. 2004. Effects of disease type and latency on the value of mortality risk. J. Risk Uncertain. 28(1):73-95.

Hansen, D.A., E. Edgerton, B. Hartsell, J. Jansen, H. Burge, P. Koutrakis, C. Rogers, H. Suh, J. Chow, B. Zielinska, P. McMurry, J. Mulholland, A. Russell, and R. Rasmussen. 2006. Air quality measurements for the aerosol research and inhalation epidemiology study. J. Air Waste Manage. Assoc. 56(10):1445-1458.

Harsanyi, J.C. 1955. Cardinal welfare, individualistic ethics, and interpersonal comparisons of utility. J. Polit. Econ. 63(4):309-321.

Hollingsworth, J.W., S.R. Kleeberger, and W.M. Foster. 2007. Ozone and pulmonary innate immunity. Proc. Am. Thorac. Soc. 4(3):240-246.

Holz, O., R.A. Jörres, P. Timm, M. Mücke, K. Richter, S. Koschyk, and H. Magnussen. 1999. Ozone-induced airway inflammatory changes differ between individuals and are reproducible. Am. J. Respir. Crit. Care Med. 159(3):776-784.

Holz, O., R. Tal-Singer, F. Kanniess, K.J. Simpson, A. Gibson, R.S. Vessey, S. Janicki, H. Magnussen, R.A. Jörres, and K. Richter. 2005. Validation of the human ozone challenge model as a tool for assessing anti-inflammatory drugs in early development. J. Clin. Pharmacol. 45(5):498-503.

Hu, S.C., A. Ben-Jebria, and J.S. Ultman. 1994. Longitudinal distribution of ozone absorption in the lung: Effects of respiratory flow. J. Appl. Physiol. 77(2):574-583.

Hubbell, B. 2006. Implementing QALYs in the analysis of air pollution regulations. Environ. Resour. Econ. 34(3):365-384.

Hudman, R.C., D.J. Jacob, O.R. Cooper, M.J. Evans, C.L. Heald, R.J. Park, F. Fehsenfeld, F. Flocke, J. Holloway, G. Hübler, K. Kita, M. Koike, Y. Kondo, A. Neuman, J. Nowak, S. Oltmans, D. Parrish, J.M. Roberts, and T. Ryerson. 2004. Ozone production in transpacific Asian pollution plumes and implications for ozone air quality in California. J. Geophys. Res. D Atmos. 109(23):1-14.

Hurley, J., M. Holland, A. Markandya, B. Miller, H. Anderson, J. Ayres, P. Donnan, R. Harrison, K. King, J. Stedman, and K. Stevenson. 2000. Towards Assessing and Costing the Health Impacts of Ambient Particulate Air Pollution in the UK. Final Research Report. Report TM/00/07. Institute of Occupational Medicine, Edinburgh, UK.

iHAPSS (Internet-Based Health and Air Pollution Surveillance System). 2005. Data: Mortality, Air Pollution, and Meteorological Data for 108 U.S. Cities 1987-2000. NMMAPS data [online]. Available: http://www.ihapss.jhsph.edu/data/NMMAPS/descriptives/frame.htm [accessed Mar. 27, 2008].

Ihorst, G., T. Frischer, F. Horak, M. Schumacher, M. Kopp, J. Forster, J. Mattes, and J. Kuehr. 2004. Long- and medium-term ozone effects on lung growth including a broad spectrum of exposure. Eur. Respir. J. 23(2):292-299.

IOM (Institute of Medicine). 2006. Valuing Health for Regulatory Cost-Effectiveness Analysis. Washington, DC: The National Academies Press.

IPCC (Intergovernmental Panel on Climate Change). 2001. Tropospheric O_3. Pp. 260-263 in Climate Change 2001: The Scientific Basis. Cambridge, UK: Cambridge University Press [online]. Available: http://www.grida.no/climate/ipcc_tar/wg1/142.htm [accessed Oct. 31, 2007].

Ito, K., S.F. De Leon, and M. Lippman. 2005. Associations between ozone and daily mortality: A review and an additional analysis. Epidemiology 16(4):446-457.

Jacob, D.J., J.A. Logan, and P.P. Murti. 1999. Effect of rising Asian emissions on surface ozone in the United States. Geophys. Res. Lett. 26(14):2175-2178.

Jacobson, M.Z. 1998. Studying the effects of aerosols on vertical photolysis rate coefficient and temperature profiles over an urban airshed. J. Geophys. Res. 103(D9):10593-10604.

Jenkins, H.S., J.L. Devalia, R.L. Mister, A.M. Bevan, C. Rusznak, and R.J. Davies. 1999. The effect of exposure to ozone and nitrogen dioxide on the airway response of atopic asthmatics to inhaled allergen: Dose- and time-dependent effects. Am. J. Respir. Crit. Care Med. 160(1):33-39.

Joad, J.P., K.S. Kott, J.M. Bric, J.L. Peake, C.G. Plopper, E.S. Schelegle, L.J. Gershwin, and K.E. Pinkerton. 2006. Structural and functional localization of airway effects from episodic exposure of infant monkeys to allergen and/or ozone. Toxicol. Appl. Pharmacol. 214(3):237-243.

Johannesson, M., and P.O. Johansson. 1996. To be, or not to be, that is the question: An empirical study of the WTP for an increased life expectancy at an advanced age. J. Risk Uncertain. 13(2):163-174.

Johannesson, M., P.O. Johansson, and K.G. Löfgren. 1997. On the value of changes in life expectancy: Blips versus parametric changes. J. Risk Uncertain. 15(3):221-239.

Johnson, F.R., W.H. Desvousges, M.C. Ruby, D. Stieb, and P. De Civita. 1998. Eliciting stated health preferences: An application to willingness to pay for longevity. Med. Decis. Making 18(2):S57-S67.

Jones-Lee, M.W. 1989. The Economics of Safety and Physical Risk. Oxford: Blackwell.

Jones-Lee, M.W. 1991. Altruism and the value of other people's safety. J. Risk Uncertain. 4(2):213-219.

Jones-Lee, M.W., M. Hammerton, and P.R. Philips. 1985. The value of safety: Results of a National Sample Survey. Econ. J. 95(March):49-72.

Jones-Lee, M.W., G. Loomis, D. O'Reilly, and P.R. Phillips. 1993. The Value of Preventing Nonfatal Road Injuries: Findings of a Willingness-to-Pay National Sample Survey. Working Paper WP/ SRC/2. Transport Research Laboratory, Crowthorn.

Judek, S., B. Jessiman, D. Stieb, and R. Vet. 2005. Estimated Number of Excess Deaths in Canada Due to Air Pollution. Health Canada and Environment Canada [online]. Available: http://www.hc-sc.gc.ca/ahc-asc/media/nr-cp/2005/2005_32bk2_e.html [accessed Jan 2, 2008].

Kabel, J.R., A. Ben-Jebria, and J. Ultman. 1994. Longitudinal distribution of ozone absorption in the lung: Comparison of nasal and oral quiet breathing. J. Appl. Physiol. 77(6):2584-2592.

Kajekar, R., E.M. Pieczarka, S.M. Smiley-Jewell, E.S. Schelegle, M.V. Fanucchi, and C.G. Plopper. 2007. Early postnatal exposure to allergen and ozone leads to hyperinnervation of the pulmonary epithelium. Respir. Physiol. Neurobiol. 155(1):55-63.

Kelsall, J.E., S.L. Zeger, and J.M. Samet. 1999. Frequency domain Log-linear models; Air pollution and mortality. Appl. Stat. 48(3):332-344.

Kleeberger, S.R. 1995. Genetic susceptibility to ozone. Toxicol. Lett. 82-83:295-300.

Kleeberger, S.R. 2005. Genetic aspects of pulmonary responses to inhaled pollutants. Exp. Toxicol. Pathol. 57(Suppl. 1):147-153.

Kleeberger, S.R., R.C. Levitt, L.Y. Zhang, M. Longphre, J. Harkema, A. Jedlicka, S.M. Eleff, D. DiSilvestre, and K.J. Holroyd. 1997. Linkage analysis of susceptibility to ozone-induced lung inflammation in inbred mice. Nat. Genet. 17(4):475-478.

Klepeis, N.E.,W.C Nelson, W.R. Ott, J.P. Robinson, M. Tsang, P. Switzer, J.V. Behar, S.C. Hern, and W.H. Engelmann. 2001. The National Human Activity Pattern Survey (NHAPS): A resource for assessing exposure to environmental pollutants. J. Expo. Anal. Environ. Epidemiol. 11(3):231-252.

Kochi, I., B. Hubbell, and R. Kramer. 2006. An empirical Bayes approach to combining and comparing estimates of the value of statistical life for environmental policy analysis. Environ. Resour. Econ. 34(3):385-406.

Koren, H.S., R.B. Devlin, D.E. Graham, R. Mann, M.P. McGee, D.H. Horstman, W.J. Kozumbo, S. Becker, D.E. House, W.F. McDonnell, and P.A. Bromberg. 1989. Ozone-induced inflammation in the lower airways of human subjects. Am. Rev. Respir. Dis. 139(2):407-415.

Koren, H.S., R.B. Devlin, S. Becker, R. Perez, and W.F. McDonnell. 1991. Time-dependent changes of markers associated with inflammation in the lungs of humans exposed to ambient levels of ozone. Toxicol. Pathol. 19(4 Pt 1):406-411.

Krewski, D., R.T. Burnett, M.S. Goldberg, K. Hoover, J. Siemiatycki, M. Jerrett, M. Abrahamowicz, and W.H. White. 2000. Reanalysis of the Harvard Six Cities Study and the American Cancer Society Study of Particulate Air Pollution and Mortality: A Special Report of the Institute's Particulate Epidemiology Reanalysis Project. Cambridge, MA: Health Effects Institute.

Krupnick, A. 2007. Mortality-risk valuation and age: Stated preference evidence. Rev. Environ. Econ. Policy 1(2):261-282.

Krupnick, A., A. Alberini, M. Cropper, N. Simon, B. O'Brien, R. Goeree, and M. Heintzelman. 2002. Age, health and the willingness to pay for mortality risk reductions: A contingent valuation survey of Ontario residents. J. Risk Uncertain. 24(2):161-186.

Künzli, N., F. Lurmann, M. Segal, L. Ngo, J. Balmes, and I.B. Tager. 1997. Association between lifetime ambient ozone exposure and pulmonary function in college freshmen-results of a pilot study. Environ. Res. 72(1):8-23.

Laden, F., L.M. Neas, D.W. Dockery, and J. Schwartz. 2000. Association of fine particulate matter from different sources with daily mortality in six U.S. cities. Environ. Health Perspect. 108(10):941-947.

Lefohn, A.S. 2006. Presentation to the Clean Air Science Advisory Committee Ozone Panel, August 24, 2006.

Levy, J.I., T.J. Carrothers, J.T. Tuomisto, J.K. Hammitt, and J.S. Evans. 2001. Assessing the public health benefits of reduced ozone concentrations. Environ. Health Perspect. 109(12):1215-1216.

Levy, J.I., S.M. Chemerynski, and J.A. Sarnat. 2005. Ozone exposure and mortality risk: An empirical Bayes Meta-Regression Analysis. Epidemiology 16(4):458-468.

Liao, K.J., E. Tagaris, K. Manomaiphiboon, S.L. Napelenok, J.H. Woo, S. He, P. Amar, and A. Russell. 2007. Sensitivities of ozone and fine particulate matter formation to emissions under the impact of potential future climate change. Environ. Sci. Technol. 41(24):8355-8361.

Liao, K.J., E. Tagaris, S.L. Napelenok, K. Manomaiphiboon, J.H. Woo, P. Amar, S. He, and A.G. Russell. 2008. Current and future linked responses of ozone and PM2.5 to emissions controls. Environ. Sci. Technol. 42(13):4670-4675.

Linaker, C.H., A.J. Chauhan, H.M. Inskip, S.T. Holgate, and d. Coggon. 2000. Personal exposures of children to nitrogen dioxide relative to concentrations in outdoor air. Occup. Environ. Med. 57(7):472-476.

Liu, J.T., and J.K. Hammitt. 1999. Perceived risk and value of workplace safety in a developing country. J. Risk Res. 2(3):263-275.

Liu, L.J.S., P. Koutrakis, H.H. Suh, J.D. Mulik, and R.M. Burton. 1993. Use of personal measurements for ozone exposure assessment: A pilot-study. Environ. Health Perspect. 101(4):318-324.

Liu, L.J., R. Delfino, and P. Koutrakis. 1997. Ozone exposure assessment in a southern California community. Environ Health Perspect. 105(1):58-65.

Ludwig, J., and P.J. Cook. 2001. The benefits of reducing gun violence: Evidence from contingent valuation survey data. J. Risk Uncertain. 22(3):207-226.

Magat, W.A., W.K. Viscusi, and J. Huber. 1996. A reference lottery metric for valuing health. Manage. Sci. 42(8):1118-1130.

Marcus, A.H., and S.R. Kegler. 2001. Confounding in air pollution epidemiology: When does two-stage regression identify the problem? Environ. Health Perspect. 109(12):1193-1196.

Mauskopf, J.A., and M.T. French. 1991. Estimating the value of avoiding morbidity and mortality from foodborne illnesses. Risk Anal. 11(4):619-631.

Maynard, R. 2004. Key airborne pollutants – the impact on health. Sci. Total Environ. 334-335:9-13.

McConnell, R., K. Berhane, F. Gilliland, S.J. London, T. Islam, W.J. Gauderman, E. Avol, H.G. Margolis, and J.M. Peters. 2002. Asthma in exercising children exposed to ozone: A cohort study. Lancet 359(9304):386-391.

McConnell, R., K. Berhane, L. Yao, F.W. Lurmann, E. Avol, and J.M. Peters. 2006. Predicting residential ozone deficits from nearby traffic. Sci Total Environ. 363(1-3):166-174.
McCunney, R.J. 2005. Asthma, genes, and air pollution. J. Occup. Environ. Med. 47(12):1285-1291.
McCurdy, T., G. Glen, L. Smith, and Y. Lakkadi. 2000. The national exposure research laboratory's consolidated human activity database. J. Expo. Anal. Environ. Epidemiol. 10(6 Pt 1):566-578.
McDonnell, W.F. 1991. Intersubject variability in human acute ozone responsiveness. Pharmacogenetics 1(2):110-113.
Medina-Ramon, M., and J. Schwartz. In press. Who is more vulnerable to die from ozone air pollution? Epidemiology
Miller, T.R. 1990. The plausible range for the value of life—Red herrings among the mackerel. J. Forensic Econ. 3(3):17-40.
Miller, B.G, and B. Armstrong. 2001. Quantification of the Impacts of Air Pollution on Chronic Cause-Specific Mortality. Research Report TM/01/08. Edinburgh: Institute of Occupational Medicine [online]. Available: http://www.iom-world.com/pubs/IOM_TM0108.pdf [accessed May 2, 2008].
Miller, B.G., and J.F. Hurley. 2003. Life table methods for quantitative impact assessments in chronic mortality. J. Epidemiol. Community Health 57(3):200-206.
Miller, B.G., and J.F. Hurley. 2006. Comparing Estimated Risks for Air Pollution with Risks for Other Health Effects. Research Report TM/06/01. Edinburgh: Institute of Occupational Medicine. March 2006 [online]. Available: http://www.iom-world.org/pubs/IOM_TM0601.pdf [accessed Mar. 28, 2008].
Miller, K.A., D.S. Siscovick, L. Sheppard, K. Shepherd, J.H. Sullivan, G.L. Anderson, and J.D. Kaufman. 2007. Long-term exposure to air pollution and incidence of cardiovascular events in women. N. Engl. J. Med. 356(5):447-458.
Mills, N.L., H. Törnqvist, M.C. Gonzalez, E. Vink, S.D. Robinson, S. Söderberg, N.A. Boon, K. Donaldson, T. Sandström, A. Blomberg, and D.E. Newby. 2007. Ischemic and thrombotic effects of dilute diesel-exhaust inhalation in men with coronary heart disease. N. Engl. J. Med. 357(11):1075-1082.
Mishan, E.J. 1971. Evaluation of life and limb: A theoretical approach. J. Polit. Econ. 79(4):687-705.
Mitchell, R.C., and R.T. Carson. 1989. Using Surveys to Value Public Goods: The Contingent Valuation Method. Washington, DC: Resources for the Future.
Morris, J., and J.K. Hammitt. 2001. Using life expectancy to communicate benefits of health care programs in contingent valuation studies. Med. Decis. Making 21(6):468-478.
Mrozek, J.R., and L.O. Taylor. 2002. What determines the value of life? A meta-analysis. J. Policy Anal. Manage. 21(2):253-270.
Murray, C.J., and C.R. Nelson. 2000. State-space modeling of the relationship between air quality and mortality. J. Air Waste Manage. Assoc. 50(7):1075-1080.
NAE (National Academy of Engineering). 1972. Perspectives on Benefit-Risk Decision Making. Washington, DC: National Academy of Engineering.
Napelenok, S.L., D.S. Cohan, Y. Hu, and A.G. Russell. 2007. Decoupled direct 3D sensitivity analysis method (DDM-3D/PM). Atmos. Environ. 40(32):6112-6121.
NARSTO. 2003. Particulate Matter Science for Policy Makers, A NARSTO Assessment. EPRI 1007735. Palo Alto, CA: EPRI. February 2003.

NARSTO. 2004. Particulate Matter Assessment for Policy Makers: A NARSTO Assessment, P. McMurry, M. Shepherd, and J. Vickery, eds. Cambridge, UK: Cambridge University Press.

Neill, J. 2006. Meta-Analysis Research Methodology [online]. Available: http://wilderdom.com/research/meta-analysis.html [accessed Dec. 5, 2007].

Nightingale, J.A., D.F. Rogers, K.F. Chung, and P.J. Barnes. 2000. No effect of inhaled budesonide on the response to inhaled ozone in normal subjects. Am. J. Respir. Crit. Care Med. 161(2 Pt 1):479-486.

NRC (National Research Council). 1975. Air Quality and Stationary Source Emission Control: A Report. Serial No. 94-4. Washington, DC: U.S. Government Printing Office.

NRC (National Research Council). 1982. Risk and Decision Making: Perspectives and Research. Washington, DC: National Academy Press.

NRC (National Research Council). 1983. Risk Assessment in the Federal Government: Managing the Process. Washington, DC: National Academy Press.

NRC (National Research Council). 1991. Rethinking the Ozone Problem in Urban and Regional Air Pollution. Washington, DC: National Academy Press.

NRC (National Research Council). 1994. Science and Judgment in Risk Assessment. Washington, DC: National Academy Press.

NRC (National Research Council). 1996. Understanding Risk: Informing Decisions in a Democratic Society, P.C. Stern, and H.V. Fineberg, eds. Washington, DC: National Academy Press.

NRC (National Research Council). 2002. Estimating the Public Health Benefits of Proposed Air Pollution Regulations. Washington, DC: The National Academies Press.

NRC (National Research Council). 2004a. Air Quality Management in the United States. Washington DC: The National Academies Press.

NRC (National Research Council). 2004b. Research Priorities for Airborne Particulate Matter. IV. Continuing Research Progress. Washington DC: The National Academies Press.

NRC (National Research Council). 2007a. Scientific Review of the Proposed Risk Assessment Bulletin from the Office of Management and Budget. Washington, DC: National Academies Press.

NRC (National Research Council). 2007b. Models in Environmental Regulatory Decision Making. Washington, DC: National Academies Press.

OMB (U.S. Office of Management and Budget). 2003. Regulatory Analysis. Circular A-4 to the Heads of Executive Agencies and Establishments, September 17, 2003 [online]. Available: http://www.whitehouse.gov/omb/circulars/a004/a-4.pdf [accessed Nov. 19, 2007].

OMB (U.S. Office of Management and Budget). 2005. Validating Regulatory Analysis: 2005 Report to Congress on the Costs and Benefits of Federal Regulations and Unfunded Mandates on State, Local, and Tribal Entities. U.S. Office of Management and Budget, Office of Information and Regulatory Affairs [online]. Available: http://www.whitehouse.gov/omb/inforeg/2005_cb/final_2005_cb_report.pdf [accessed Jan. 2, 2008].

OMB (U.S. Office of Management and Budget). 2006. Draft 2006 Report to Congress on the Costs and Benefits of Federal Regulations. U.S. Office of Management and Budget, Office of Information and Regulatory Affairs [online]. Available: http://www.whitehouse.gov/omb/inforeg/reports/2006_draft_cost_benefit_report.pdf [accessed Jan. 2, 2008].

O'Neill, M.S., M. Jerrett, I. Kawachi, J.I. Levy, A.J. Cohen, N. Gouveia, P. Wilkinson, T. Fletcher, L. Cifuentes, and J. Schwartz. 2003. Health, wealth, and air pollution: Advancing theory and methods. Environ. Health Perspect. 111(16):1861-1870.

Ostro, B.D., H. Tran, and J.I. Levy. 2006. The health benefits of reduced tropospheric ozone in California. J. Air Waste Manage. Assoc. 56(7):1007-1021.

Patterson, E., and D.J. Eatough. 2000. Indoor/outdoor relationships for ambient PM2.5 and associated pollutants: Epidemiological implications in Lindon, Utah. J. Air Waste Manage. Assoc. 50(1):103-110.

PCCRARM (Presidential/Congressional Commission on Risk Assessment and Risk Management). 1997. Framework for Environmental Health Risk Management: Final Report. Washington, DC: The Commission.

Peel, J.L., P.E. Tolbert, M. Klein, K.B. Metzger, W.D. Flanders, K. Todd, J.A. Mulholland, P.B. Ryan, and H. Frumkin. 2005. Ambient air pollution and respiratory emergency department visits. Epidemiology 16(2):164-174.

Peters, E.A., J.T. Hiltermann, and J. Stolk. 2001. Effect of apocynin on ozone-induced airway hyperresponsiveness to methacholine in asthmatics. Free Radic. Biol. Med. 31(11):1442-1447.

Peters, J.M., E. Avol, W.J. Gauderman, W.S. Linn, W. Navidi, S.J. London, H. Margolis, E. Rappaport, H. Vora, H. Gong Jr., and D.C. Thomas. 1999. A study of twelve Southern California communities with differing levels and types of air pollution. II. Effects on pulmonary function. Am. J. Respir. Crit. Care Med. 159(3):768-775.

Peterson, M.R., W.F. Gutknecht, R.L. Perkins, R.K.M. Jayanty, and E.D. Hardison. 2000. Laboratory support for chemical speciation of PM2.5. EM 8:17-22.

Plopper, C.G., S.M. Smiley-Jewell, L.A. Miller, M.V. Fanucchi, M.J. Evans, A.R. Buckpitt, M. Avdalovic, L.J. Gershwin, J.P. Joad, R. Kajekar, S. Larson, K.E. Pinkerton, L.S. Van Winkle, E.S. Schelegle, E.M. Pieczarka, R. Wu, and D.M. Hyde. 2007. Asthma/allergic airways disease: Does postnatal exposure to environmental toxicants promote airway pathobiology? Toxicol. Pathol. 35(1):97-110.

Pope, C.A., III, M.J. Thun, M.M. Namboodiri, D.W. Dockery, J.C. Evans, F.E. Speizer, and C.W. Heath, Jr. 1995. Particulate air pollution as a prediction of mortality in a prospective study of U.S. adults. Am. J. Respir. Crit. Care Med. 151(3Pt.1):669-674.

Pope, C.A., III, R.T. Burnett, M.J. Thun, E.E. Calle, D. Krewski, K. Ito, and G.D. Thurston. 2002. Lung cancer, cardiopulmonary mortality, and long-term exposure to fine particulate air pollution. JAMA 287(9):1132-1141.

Pope, C.A., III, R.T. Burnett, G.D. Thurston, M.J. Thun, E.E. Calle, D. Krewski, and J.J. Godleski. 2004. Cardiovascular mortality and long-term exposure to particulate air pollution: Epidemiological evidence of general pathophysiological pathways of disease. Circulation 109(1):71-77.

Pradhan, A.D., J.E. Manson, N. Rifai, J.E. Buring, and P.M. Ridker. 2001. C-reactive protein, interleukin 6, and risk of developing type 2 diabetes mellitus. JAMA 286(3):327-334.

Pradhan, A.D., N. Rifai, and P.M. Ridker. 2002. Soluble intercellular adhesion molecule-1, soluble vascular adhesion molecule-1, and the development of symptomatic peripheral arterial disease in men. Circulation 106(7):820-825.

Pratt, J.W., and R.J. Zeckhauser. 1996. Willingness to pay and the distribution of risk and wealth. J. Polit. Econ. 104(4):747-763.

Rabl, A. 2006. Analysis of air pollution mortality in terms of life expectancy changes: Relation between time series, intervention, and cohort studies. Environ. Health 5(1):1.

Ratto, J., H. Wong, J. Liu, J. Fahy, H. Boushey, C. Solomon, and J. Balmes. 2006. Effects of multiday exposure to ozone on airway inflammation as determined using sputum induction. Environ. Health Perspect. 114(2):209-212.

Rawls, J.A. 1971. A Theory of Justice. Cambridge, MA: Harvard University Press.

Reisenauer, C.S., J.Q. Koenig, M.S. McManus, M.S. Smith, G. Kusic, and W.E. Pierson. 1988. Pulmonary response to ozone exposures in healthy individuals aged 55 years or greater. JAPCA 38(1):51-55.

Ridker, P.M., R.J. Glynn, and C.H. Hennekens. 1998. C-reactive protein adds to the predictive value of total and HDL cholesterol in determining risk of first myocardial infarction. Circulation 97(20):2007-2011.

Ridker, P.M., C.H. Hennekens, J.E. Buring, and N. Rifai. 2000. C-reactive protein and other markers of inflammation in the prediction of cardiovascular disease in women. N. Engl. J. Med. 342(12):836-843.

Robinson, L.A. 2004. Current Federal Agency Practices for Valuing the Impact of Regulations on Human Health and Safety. Prepared for the Committee to Evaluate Measures of Health Benefits for Environmental, Health, and Safety Regulation, Institute of Medicine, Washington, DC.

Robinson, L. 2007. How U.S. government agencies value mortality risk reductions. Rev. Environ. Econ. Policy 1(2):283-299.

Robinson, A.L., N.M. Donahue, M.K. Shrivastava, E.A. Weitkamp, A.M. Sage, A.P. Grieshop, T.E. Lane, J.R. Pierce, and S.N. Pandis. 2007. Rethinking organic aerosols: Semivolatile emissions and photochemical aging. Science 315(5816):1259-1262.

Rojas-Bracho, L., H.H. Suh, P.J. Catalano, and P. Koutrakis. 2004. Personal exposures to particles and their relationships with personal activities for chronic obstructive pulmonary disease patients living in Boston. J. Air Waste Manage. Assoc. 54(2):207-217.

Rojas-Martinez, R., R. Perez-Padilla, G. Olaiz-Fernandez, L. Mendoza-Alvarado, H. Moreno-Macias, T. Fortoul, W. McDonnell, D. Loomis, and I. Romieu. 2007. Lung function growth in children with long-term exposure to air pollutants in Mexico City. Am. J. Respir. Crit. Care Med. 176(4):377-384.

Romieu, I., F. Menesses, S. Ruiz, J. Huerta, J.J. Sienra, M. White, R. Etzel, and M. Hernandez. 1997. Effects of intermittent ozone exposure on peak expiratory flow and respiratory symptoms among asthmatic children in Mexico City. Arch. Environ. Health 52(5):368-376.

Romieu, I., J.J. Sienra-Monge, M. Ramirez-Aguilar, M.M. Tellez-Rojo, H. Moreno-Marcias, N.I. Reyes-Ruiz, B.E. del Río-Navarro, M.X. Ruiz-Navarro, G. Hatch, R. Slade, and M. Hernandex-Avila. 2002. Antioxidant supplementation and lung functions among children with asthma exposed to high levels of air pollutants. Am. J. Respir. Crit. Care Med. 166(5):703-709.

Romieu, I., J.J. Sienra-Monge, M. Ramirez-Aguilar, H. Moreno-Marcias, N.I. Reyes-Ruiz, B. Estela del Río-Navarro, M. Hernández-Avila, and S.J. London. 2004. Genetic polymorphisms of GSTM1 and antioxidant supplementation influence lung function in relation to ozone exposure in asthmatic children in Mexico City. Thorax 59(1):8-10.

Romieu, I., F. Castro-Giner, N. Künzli, and J. Sunyer. 2008. Air pollution, oxidative stress and dietary supplementation: A review. Eur. Respir. J. 31(1):179-197.

Röösli, M., N. Künzli, C. Braun-Fahrländer, and M. Egger. 2005. Years of life lost attributable to air pollution in Switzerland: Dynamic exposure-response model. Int. J. Epidemiol. 34(5):1029-1035.

Russell, A.G., and G.R. Cass. 1986. Verification of a mathematical-model for aerosol nitrate and nitric-acid formation and its use for control measure evaluation. Atmos. Environ. 20(10):2011-2025.

Ryerson, T.B., M. Trainer, J.S. Holloway, D.D. Parrish, L.G. Huey, D.T. Sueper, G.J. Frost, S.G. Donnelly, S. Schauffler, E.L. Atlas, W.C. Kuster, P.D. Goldan, G. Hubler, J.F. Meagher, and F.C. Fehsenfeld. 2001. Observations of ozone formation in power plant plumes and implications for ozone control strategies. Science 292(5517):719-723.

Sardar, S.B., P.M. Fine, H. Yoon, and C. Sioutas. 2004. Associations between particle number and gaseous co-pollutant concentrations in the Los Angeles basin. J. Air Waste Manage. Assoc. 54(8):992-1005.

Sarnat, J., P. Koutrakis, and H.H. Suh. 2000. Assessing the relationship between personal particulate and gaseous exposures of senior citizens living in Baltimore, MD. J. Air Waste Manage. Assoc. 50(7):1184-1198.

Sarnat, J.A., J. Schwartz, P.J. Catalano, and H.H. Suh. 2001. Gaseous pollutants in particulate matter epidemiology: Confounders or surrogates? Environ. Health Perspect. 109(10):1053-1061.

Sarnat, J.A., K.W. Brown, J. Schwartz, B.A. Coull, and P. Koutrakis. 2005. Ambient gas concentrations and personal particulate matter exposures: Implications for studying the health effects of particles. Epidemiology 16(3):385-395.

Sarnat, J.A., A. Marmur, M. Klein, E. Kim, A.G. Russell, S.E. Sarnat, J.A. Mulholland, P.K. Hopke, and P.E. Tolbert. 2008. Fine particle sources and cardiorespiratory morbidity: An application of chemical mass balance and factor analytical source-apportionment methods. Environ. Health Perspect. 116(3):459-466.

Sarnat, S.E., B.A. Coull, J. Schwartz, D.R. Gold, and H.H. Suh. 2006. Factors affecting the association between ambient concentrations and personal exposures to particles and gases. Environ. Health Perspect. 114(5):649-654.

Savov, J.D., G.S. Whitehead, J. Wang, G. Liao, J. Usuka, G. Peltz, W.M. Foster, and D.A. Schwartz. 2004. Ozone-induced acute pulmonary injury in inbred mouse strains. Am. J. Respir. Cell Mol. Biol. 31(1):69-77.

Schelling, T.C. 1968. The life you save may be your own. Pp. 127-162 in Problems in Public Expenditure Analysis, S.B. Chase, ed. Washington, DC: The Brookings Institution.

Schwartz, J. 2000. Harvesting and long-term exposure effects in the relationship between air pollution and mortality. Am. J. Epidemiol. 151(5):440-448.

Schwartz, J. 2004. The effects of particulate air pollution on daily deaths: A multi-city case crossover analysis. Occup. Environ. Med. 61(12):956-961.

Schwartz, J. 2005. How sensitive is the association between ozone and daily deaths to control for temperature? Am. J. Respir. Crit. Care Med. 171(6):627-631.

Seal, E. Jr., W.F. McDonnell, D.E. House, S.A. Salaam, P.J. Dewitt, S.O. Butler, J. Green, and L. Raggio. 1993. The pulmonary response of white and black adults to six concentrations of ozone. Am. Rev. Respir. Dis. 147(4):804-810.

Seelye, K.Q., and J. Tierney. 2003. 'Senior Death Discount' assailed. New York Times, May 8, 2003 [online]. Available: http://www.commondreams.org/headlines03/0508-09.htm [accessed August 15, 2007].

Selgrade, M.K., C.G. Plopper, M.I. Gilmour, R.B. Conolly, and B.S. Foos. 2008. Assessing the health effects and risks associated with children's inhalation exposures-asthma and allergy. J. Toxicol. Environ. Health A. 71(3):196-207.

Shepard, D.S., and R.J. Zeckhauser. 1982. Life cycle consumption and willingness to pay for increased survival. Pp. 95-141 in Value of Life and Safety, M.W. Jones-Lee, ed. Amsterdam: North-Holland.

Sillman, S., J.A. Logan, and S.C. Wofsy. 1990. The sensitivity of ozone to nitrogen-oxides and hydrocarbons in regional ozone episodes. J. Geophys. Res. 95(D2):1837-1851.

Sisler, J.F., and W.C. Malm. 2000. Interpretation of trends of PM2.5 and reconstructed visibility from the IMPROVE network. J. Air Waste Manage. Assoc. 50(5):775-789.

Slovic, P. 1987. Perception of risk. Science 236(4799):280-285.

Stieb, D.M., S. Judek, and R.T. Burnett. 2002. Meta-analysis of time-series studies of air pollution and mortality: Effects of gases and particles and the influence of cause of death, age, and season. J. Air Waste Manage. Assoc. 52(4):470-484.

Stieb, D.M., S. Judek, and R.T. Burnett. 2003. Meta-analysis of time-series studies of air pollution and mortality: Update in relation to the use of generalized additive models. J. Air Waste Manage. Assoc. 53(3):258-261.

Sunstein, C.R. 2002. Risk and Reason: Safety, Law, and the Environment. New York: Cambridge University Press.

Tager, I.B., J. Balmes, F. Lurmann, L. Ngo, S. Alcorn, and N. Kunzli. 2005. Chronic exposure to ambient ozone and lung function in young adults. Epidemiology 16(6):751-759.

Tashkin, D.P., R. Detels, M. Simmons, H. Liu, A.H. Coulson, J. Sayre, and S. Rokaw. 1994. The UCLA population studies of chronic obstructive respiratory disease: XI. Impact of air pollution and smoking on annual change in forced expiratory volume in one second. Am. J. Respir. Crit. Care Med. 149(5):1209-1217.

Thurston, G.D., and K. Ito. 2001.Epidemiological studies of acute ozone exposures and mortality. J. Expo. Anal. Environ. Epidemiol. 11(4):286-294.

Thurston, G., K. Ito, T. Mar, W.F. Christensen, D.J. Eatough, R.C. Henry, E. Kim, F. Laden, R. Lall, T.V. Larson, H. Liu, L. Neas, J. Pinto, M. Stolzel, H. Suh, and P.K. Hopke. 2005. Results and implications of the workshop on the source apportionment of PM health effects. Epidemiology 16(5):S134-S135.

Trasande, L., and G.D. Thurston. 2005. Role of air pollution in asthma and other pediatric morbidities. J. Allergy Clin. Immunol. 115(4):689-699.

Tsevat, J., N.V. Dawson, A.W. Wu, J. Lynn, J.R. Soukup, E.F. Cook, H. Vidaillet, and R.S. Phillips. 1998. Health values of hospitalized patients 80 years and older. JAMA 279(5):371-375.

Vagaggini, B., M. Taccola, I. Conti, S. Carnevali, S. Cianchetti, M.L. Bartoli, E. Bacci, F.L. Dente, A. Di Franco, D. Giannini, and P.L. Paggiaro. 2001. Budesonide reduces neutrophilic but not functional airway response to ozone in mild asthmatics. Am. J. Respir. Crit. Care Med. 164(12):2172-2176.

Viscusi, W.K. 1993. The value of risks to life and health. J. Econ. Lit. 31(4):1912-1946.

Viscusi, W.K., and J.E. Aldy. 2003. The value of statistical life: A critical review of market estimates throughout the world. J. Risk Uncertain. 27(1):5-76.

Viscusi, W.K., and J.E. Aldy. 2007. Labor market estimates of the senior discount for the value of statistical life. J. Environ. Econ. Manage. 53(3):377-392.

Viscusi, W.K., and M.J. Moore. 1989. Rates of time preference and valuations of the duration of life. J. Public Econ. 38(3):297-317.

Viscusi, W.K., W.A. Magat, and J. Huber. 1991. Pricing environmental health risks: Survey assessment of risk-risk and risk-dollar trade-offs for chronic bronchitis. J. Environ. Econ. Manage. 21(1):32-51.

Volz, A., and D. Kley. 1988. Evaluation of the Montsouris series of ozone measurements made in the nineteenth century. Nature 332(6161):240-242.

Wade, K.S., J.A. Mulholland, A. Marmur, A.G. Russell, B. Hartsell, E. Edgerton, M. Klein, L. Waller, J.L. Peel, and P.E. Tolbert. 2006. Effects of instrument precision and spatial variability on the assessment of the temporal variation of ambient air pollution in Atlanta, Georgia. J. Air Waste Manage. Assoc. 56(6):876-888.

Wakefield, J., and G. Shaddick. 2006. Health-exposure modeling and the ecological fallacy. Biostatistics 7(3):438-455.

Waldman, J., P.Lioy, G. Thurston, and M. Lippmann. 1990. Spatial and temporal patterns in summertime sulfate aerosol acidity and neutralization within a metropolitan area. Atmos. Environ. 24B:115-126.

Weinmann, G.G., M.C. Liu, D. Proud, M. Weidenbach-Gerbase, W. Hubbard, and R. Frank. 1995. Ozone exposure in humans: Inflammatory, small and peripheral airway responses. Am. J. Respir. Crit. Care Med. 152(4 Pt 1):1175-1182.

Weschler, C.J. 2000. Ozone in indoor environments: Concentration and chemistry. Indoor Air 10(4):269-288.

Weschler, C.J. 2006. Ozone's impact on public health: Contributions from indoor exposures to ozone and products of ozone-initiated chemistry. Environ. Health Perspect. 114(10):1489-1496.

Weschler, C. J., and H.C. Shields. 1994. Indoor chemistry involving O_3, NO, and NO_2 as evidenced by 14 months of measurements at a site in Southern California. Environ. Sci. Technol. 28(12):2120-2132.

White, M.C., R.A. Etzel, W.D. Wilcox, and C. Lloyd. 1994. Exacerbations of childhood asthma and ozone pollution in Atlanta. Environ. Res. 65(1):56-68.

WHO Working Group. 2000. Evaluation and use of epidemiological evidence for environmental health risk assessment: WHO guideline document. Environ. Health Perspect. 108(10):997-1002.

WHO (World Health Organization). 2006. Air Quality Guidelines-Global Update 2005: Particulate Matter, Ozone, Nitrogen Dioxide and Sulfur Dioxide. Geneva: World Health Organization.

Wilson, A.M., J.C. Salloway, C.P. Wake, and T. Kelly. 2004. Air pollution and the demand for hospital services: A review. Environ. Int. 30(8):1109-1118.

Woo, J.H., S. He, P. Amar, E. Tagaris, K. Manomaiphiboon, K.J. Liao, and A.G. Russell. 2006. Development of Mid-Century Anthropogenic Emissions Inventory in Support of Regional Air Quality Modeling under Influence of Climate Change. Presentation at 15th International Emission Inventory Conference "Reinventing Inventories- New Ideas in New Orleans, May 15-18, 2006, New Orleans, LA [online]. Available: http://www.epa.gov/ttn/chief/conference/ei15/session4/woo2.pdf [accessed Jan. 2, 2008].

Zanobetti, A., and J. Schwartz. 2008. Mortality displacement in the association of ozone with mortality: An analysis of 48 U.S. cities. Am. J. Respir. Crit. Care Med. 177(2):184-189.

Zanobetti, A., M.P. Wand, J. Schwartz, and L.M. Ryan. 2000. Generalized additive distributed lag models: Quantifying mortality displacement. Biostatistics 1(3):279-292.

Zeger, S.L., F. Dominici, and J.M. Samet. 1999. Harvesting-resistant estimates of pollution effects on mortality. Epidemiology 10(2):171-175.

Zeger, S.L., D. Thomas, F. Dominici, J.M. Samet, J. Schwartz, D. Dockery, and A. Cohen. 2000. Exposure measurement error in time-series studies of air pollution: Concepts and consequences. Environ. Health Perspect. 108(5):419-426.

Zhang, J.F., and P.J. Lioy. 1994. Ozone in residential air: Concentrations, I/O ratios, indoor chemistry, and exposures. Indoor Air 4(2):95-105.

Zhu, Y.F., W.C. Hinds, S. Kim, S. Shen, and C. Sioutas. 2002. Study of ultrafine particles near a major highway with heavy-duty diesel traffic. Atmos. Environ. 36(27):4323-4335.

Abbreviations

ACS:	American Cancer Society
APEX:	Air Pollutants Exposure
CAA:	U.S. Clean Air Act
CAIR:	Clean Air Implementation Rule
CASAC:	Clean Air Scientific Advisory Committee (of EPA)
CDC:	Centers for Disease Control and Prevention (of DHHS)
CHAD:	Consolidated Human Activity Database
CMS:	Centers for Medicare and Medicaid Services (of DHHS)
CMAQ:	Community Multiscale Air Quality
DHHS:	U.S. Department of Health and Human Services
DOT:	U.S. Department of Transportation
EC:	elemental carbon
EPA:	U.S. Environmental Protection Agency
ExternE:	European Union program to estimate the mortality impacts of exposure to ozone and to value those impacts.
FDA:	U.S. Food and Drug Administration (of DHHS)
FMCSA:	Federal Motor Carrier Safety Administration (of DOT)
NAAQS:	National Ambient Air Quality Standard
NHTSA:	National Highway Traffic Safety Administration (of DOT)
NO:	nitric oxide
NO_2:	nitrogen dioxide
NO_x:	oxides of nitrogen (NO and NO_2)
O_2:	diatomic oxygen
OH:	hydroxyl radical
OP:	oxygenated organic product
OMB:	U.S. Office of Management and Budget
PM:	particulate matter
$PM_{2.5}$:	particulate matter with aerodynamic equivalent diameter of no more than 2.5 μm (microns)
PM_{10}:	particulate matter with aerodynamic equivalent diameter of no more than 10 μm
RIA:	regulatory impacts analysis

QALY:	quality-adjusted life-years
ROS:	reactive oxygen species
SAB:	Science Advisory Board (of EPA)
SIP:	state implementation plan
SO_2:	sulfur dioxide
SOA:	secondary organic aerosol
VOC:	volatile organic compound
VSL:	value of a statistical life
VSLY:	value of a statistical life-year
WTA:	willingness to accept compensation
WTP:	willingness to pay

Appendix A

Biographic Information on Committee on Estimating Mortality Risk Reduction Benefits from Decreasing Tropospheric Ozone Exposure

John C. Bailar III *(Chair)* is professor emeritus in the Department of Health Studies at the University of Chicago and scholar-in-residence at the National Academies. He is a retired commissioned officer of the U.S. Public Health Service and worked for the National Cancer Institute for 22 years. He has also held academic appointments at Harvard University and McGill University. Dr. Bailar's research interests include assessing health risks posed by chemical hazards and air pollutants and interpreting statistical evidence in medicine with emphasis on cancer. He was editor-in-chief of the *Journal of the National Cancer Institute* for 6 years and statistical consultant and member of the Editorial Board of the *New England Journal of Medicine*. Dr. Bailar is a member of the International Statistical Institute and was elected to the Institute of Medicine in 1993. He served as chair of the National Research Council Committee on Estimating the Health-Risk-Reduction Benefits of Proposed Air Pollution Regulations. He received his MD from Yale University and his PhD in statistics from American University.

Richard T. Burnett is a senior research scientist with the Healthy Environments and Consumer Safety Branch of Health Canada, where he has been working since 1983 on issues related to the health effects of outdoor air pollution. Dr. Burnett's work has focused on the use of administrative health and environmental information to determine the public-health impacts of combustion-related pollution with nonlinear random-effects models and time-series and spatial analytic techniques. Dr. Burnett was a member of the National Research Council's Committee on the Management of Air Quality in the United States and the Institute of Medicine's Committee on Valuing Health for Regulatory Cost-Effectiveness Analysis. He is on the Editorial Board of *Risk Analysis*. Dr. Burnett is an adjunct professor in the Department of Epidemiology and Community Medicine of the Faculty of Medicine, an affiliate scientist of the Institute of Population Health, and a scientist with the McLaughlin Center for Population Health Risk Assessment at the Institute of Population Health, all in the University of Ottawa. Dr. Burnett received his PhD in mathematical statistics from Queen's University.

Lauraine G. Chestnut is a managing economist at Stratus Consulting Inc. Ms. Chestnut specializes in environmental and natural-resources economics, policy analysis, and survey research. She focuses on the quantification and economic valuation of human health, visibility, and other welfare effects of environmental pollutants. She has conducted original economic and survey research to estimate the value to the public of protecting human health, visibility aesthetics, and cultural materials from the effects of air pollution and has conducted epidemiologic studies of the effects of particulate matter on human health. Ms. Chestnut has synthesized the epidemiologic and economics literature on the human health effects of air pollutants and has applied this information in numerous assessments of the benefits of air-pollution control. She has published three books and several articles in peer-reviewed journals, including the *Journal of Environmental Management, Land Economics,* the *Journal of the Air Pollution Control Association*, the *Journal of Policy Analysis and Management,* and *Archives of Environmental Health.* Ms. Chestnut served 6 years on the Environmental Protection Agency Science Advisory Board's Advisory Council on Clean Air Compliance Analysis and served on California's Air Quality Advisory Committee. She was elected to the Board of Directors of the Association of Environmental and Resource Economists. She received her MA in economics from the University of Colorado.

W. Michael Foster is a research professor of medicine in the Division of Pulmonary and Critical Care Medicine at Duke University Medical Center. Dr. Foster's laboratory performs research on humans and animal models and investigates the biologic effects of inhalational hazards (particles and gases) on airway and parenchymal lung tissues. Subjects of interest and expertise include effects of oxidant-type air pollution on lung epithelial membrane physiology, in vivo functional and biochemical tissue responses of the human and animal lung, and host (genetic) factors as modulators of the pulmonary response to ambient air pollutants, inhalable irritants, and nuisance bioaerosols. Dr. Foster received a PhD in physiology and environmental science from New York University.

A. Myrick Freeman III is the William D. Shipman Professor of Economics Emeritus at Bowdoin College. During his tenure at Bowdoin, Dr. Freeman has served as chair of the Economics Department and director of the Environmental Studies Program. He also has served as the Robert M. La Follette Distinguished

Visiting Professor at the University of Wisconsin, Madison and as a senior fellow at Resources for the Future. His principal research interests are in applied welfare economics, benefit-cost analysis, and risk management as related to issues in environmental and resource management. Much of his work has been devoted to developing models and techniques for estimating the welfare effects of environmental changes (for example, the benefits of controlling pollution and the damage to natural resources due to the releases of chemicals). He has served on the National Research Council Board on Toxicology and Environmental Health Hazards. He also has served as a member of the Environmental Protection Agency Advisory Council on Clean Air Compliance Analysis, Clean Air Science Advisory Committee, Environmental Economics Advisory Committee, and Science Advisory Board. He is the author of *The Benefits of Environmental Improvement: Theory and Practice* (1979) and *The Measurement of Environmental and Resource Values: Theory and Methods* (1993), which were cited by the Association of Environmental and Resource Economists as Publications of Enduring Value in 2003. Dr. Freeman received his PhD from the University of Washington.

Montserrat Fuentes is an associate professor in the Department of Statistics at North Carolina State University (NCSU) and holds associate status in the NCSU Department of Marine, Earth, and Atmospheric Sciences. Dr. Fuentes has developed new statistical methods that she applies to air-pollution, weather-prediction, hurricane-forecasting, and environmental health risk-assessment problems in collaboration with the air-quality modelers and scientists at the Environmental Protection Agency (EPA) and the National Center for Atmospheric Research. That work has led to numerous publications in statistical journals and books and in journals in the atmospheric sciences. She received the Abdel El-Shaarawi Young Researcher's Award in recognition of outstanding contributions to environmetric research in 2003. She is a member-elect of the International Statistical Institute and was a member of the Regional Advisory Board (Eastern North American Region) of the International Biometric Society. Dr. Fuentes is a member of the Exposure and Human Health Committee of the Environmental Protection Agency Science Advisory Board and the U.S. representative in the Board of Directors of the International Environmetrics Society. She is a member of the Biostatistical Methods and Research Design study section of the National Institutes of Health. Dr. Fuentes received her PhD in statistics from the University of Chicago.

Daniel S. Greenbaum is the president and chief executive officer of the Health Effects Institute, an independent not-for-profit research institute funded jointly by government and industry to provide trusted research on the health effects of air pollution. At the Health Effects Institute, Mr. Greenbaum has overseen the development and implementation of a research plan that focuses the institute's efforts on providing critical research on and reanalysis of particulate matter, air toxics, and alternative fuels, increasingly on an international scale. In 1999, he served as chair of the Environmental Protection Agency (EPA) Blue Ribbon Panel on Oxygenates in Gasoline, and in 2001, as chair of the EPA Clean Diesel Independent Review Panel, which affirmed that the 2007 heavy-duty diesel rules could be met. Before joining the Health Effects Institute, he served as commissioner of the Massachusetts Department of Environmental Protection. In the National Research Council, he has served on the Board on Environmental Studies and Toxicology, as vice chair of the Committee on Air Quality Management in the United States, and as a member of the Committee on Research Priorities for Airborne Particulate Matter. Mr. Greenbaum earned a master's degree in city planning from the Massachusetts Institute of Technology.

Alan Krupnick is a senior fellow and director of quality of the environment at Resources for the Future. His research focuses on analyzing environmental issues, in particular the benefits, costs, and design of air-pollution policies in the United States and in developing countries. His research also addresses the valuation of health and ecologic improvements and, more recently, the ancillary benefits of climate policy and urban transportation and development problems. Dr. Krupnick has served as a consultant to state governments, federal agencies, private corporations, the Canadian government, the European Union, the World Health Organization, and the World Bank. He co-chaired an advisory committee that counseled the U.S. Environmental Protection Agency on new ozone and particulate-matter standards. Dr. Krupnick has participated in several National Research Council studies, including those of the Committee for the Evaluation of the Congestion Mitigation and Air Quality Improvement Program, the Committee on Research and Peer Review in EPA, and the Surface Transportation Environmental Cooperative Research Program Advisory Board. He also has served on a Royal Society of Canada committee analyzing ambient-air quality standard-setting in Canada. Dr. Krupnick received his PhD in economics from the University of Maryland.

Nino Künzli is a research professor of the Center for Research in Environmental Epidemiology (CREAL) at Municipal Institute of Medical Research (IMIM), Barcelona, Spain, and associate professor in leave of absence at the University of Southern California Keck School of Medicine (Department of Preventive Medicine; Environmental Health Science Division), Los Angeles. His research focus is environmental epidemiology with an emphasis on air-pollution epidemiology, including exposure and risk assessment. He has completed a European assessment of the public-health impact of outdoor and traffic-related air pollution. He was a member of the World Health Organization Air Pollution Health Impact Assessment Working Group. He also has served on the National Research Council Committee on Estimating the Health-Risk-Reduction Benefits of Proposed Air Pollution Regulations. Dr. Künzli received his MD from the University of Basel and his MPH and PhD from the University of California, Berkeley.

Kent E. Pinkerton is a professor of pediatric medicine and anatomy, physiology, and cell biology at the University of California, Davis. He also serves as director of the university's Center for Health and the Environment. His research focuses on the health effects of environmental air pollutants on lung structure and function, the interaction of gases and airborne particles in specific sites and cell populations of the lungs in acute and chronic lung injury, and the effects of environmental tobacco smoke on lung growth and development. He received a PhD in pathology from Duke University.

Armistead Russell is a professor of environmental engineering at the Georgia Institute of Technology. His research subjects include aerosol dynamics, atmospheric chemistry, emissions control, and air pollution-control strategy design and computer modeling. Dr. Russell has served on a number of National Research Council committees and was chair of the Committee to Review EPA's Mobile Source Emissions Factor (MOBILE) Model. He received a PhD in mechanical engineering from the California Institute of Technology.

Helen Suh is an associate professor of environmental chemistry and exposure assessment at Harvard University's School of Public Health. Her research interests—air-pollutant exposures and their relationship to human health—involve the effect of exposure error, exposure modification, and confounding on epidemiology and toxicology, the contribution of indoor and outdoor sources to air-pollutant exposures, and the impact of these exposures on cardiovascular health. Her work, including research conducted at the Environmental Protection Agency Center for Ambient Particle Health Effects, has been published in numerous journals. Dr. Suh received her ScD from Harvard University's School of Public Health.

Evelyn O. Talbott is a professor in the Department of Epidemiology and a professor in the Department of Speech Communication Sciences and Disorders at the University of Pittsburgh Graduate School of Public Health. Her major focus of research is women's health and environmental epidemiology; her specific research interests include cardiovascular risk factors in women, environmental factors and cancer, and risk of coronary heart disease in women with polycystic ovary syndrome. She is the director of the University of Pittsburgh Academic Partner for Excellence in Environmental Public Health Tracking program, one of four academic centers in the United States that are funded by the Centers for Disease Control and Prevention (CDC) and charged to assist CDC implementation of a nationwide environmental public-health tracking network. She was a coinvestigator in an Allegheny County study of the relationship between air pollution and cardiopulmonary admissions of older people. Dr. Talbott received a PhD in epidemiology from the University of Pittsburgh.

Appendix B

Environmental Protection Agency's Regulatory Impact Analysis for the Final Ozone National Ambient Air Quality Standard

Table B-1 (EPA 2008b, Table 7-14) shows a variety of assumptions about the association between ozone exposure and mortality. The table is presented here to illustrate that EPA had included the assumption of no causal association between estimated reductions in the incidence of premature mortality and reductions in ozone exposure. The committee recommends that future regulatory impact analyses (RIAs) give little or no weight to that assumption unless new information that refutes the interpretation of this association as causal emerges (see Chapter 6). Presentations like that included in Table 7-14 should be revised in light of this recommendation.

TABLE B-1 (TABLE 7-14) Illustrative Strategy to Attain 0.075 ppm: Estimated Annual Reductions in the Incidence of Premature Mortality Associated with Ozone Exposure in 2020 (Incremental to Current Ozone Standard, Arithmetic Mean, 95% Confidence Intervals in Parentheses)[b, c, d, e]

Model or Assumption[a]	Reference	National Full Attainment
NMAPS	Bell et al. 2004	71 (27-110)
Meta-Analysis	Bell et al. 2005	230 (120-340)
	Ito et al. 2005	310 (200-430)
	Levy et al. 2005	320 (230-420)
Assumption that association is not causal		0

[a]Does not represent equal weighting among models or between assumption of causality vs no causality (see text in section 6.3.2.1 [of EPA 2008b]).
[b]With the exception of the assumption of no causal relationship, the arithmetic mean and 95% credible interval around the mean estimates of the annual number of lives saved are based on an assumption of a normal distribution.
[c]A credible interval is a posterior probability interval used in Bayesian statistics, which is similar to a confidence interval used in frequentist statistics.
[d]All estimates rounded to two significant figures. As such, confidence intervals may not be symmetrical.
[e]This table reflects full attainment in all locations of the U.S. except two areas of California. These two areas, which have high levels of ozone, are not planning to meet the current standard until after 2020. The estimates in the table do not reflect benefits for the San Joaquin and South Coast Air Basins.
Source: EPA 2008b.